MANCHESTER'S **METROLINK**

ALAN YEARSLEY

METRO & LIGHT RAIL SYSTEMS IN FOCUS

PLATFORM 5

© 2021 Platform 5 Publishing Ltd. All rights reserved. No part of this publication may be reproduced or transmitted in any form or by any means electronic, mechanical, photocopying, recording or otherwise, without prior permission of the publisher.
Published by Platform 5 Publishing Ltd, 52 Broadfield Road, Sheffield, S8 0XJ. England.
Printed in England by The Amadeus Press, Cleckheaton, West Yorkshire
ISBN: 978 1 909431 85 0

Front cover (top left): First generation Manchester car 1007 passes along Slade Lane having just left Stockport Road on its way to Birchfields Road depot as Manchester's official last tram on the final day of tram services in the city, 9 January 1949. *Manchester Transport Museum Society collection*

Front cover (top right): T68 car 1005 leaves Weaste and crosses Foster Street with an Eccles–Victoria service on 8 October 2008. *Robert Pritchard*

Front cover (main photo): M5000 car 3114 is seen amid the rolling countryside and farms between Shaw & Crompton and Derker on the Oldham Loop with a Rochdale–East Didsbury service on 19 April 2019. *Robert Pritchard*

Back cover: M5000 car 3036 departs from Central Park with a Rochdale–St Werburgh's Road service on 19 April 2013. *Alan Yearsley*

CONTENTS

INTRODUCTION

Welcome to the second in our series of guidebooks on UK light rail systems, following the success of our first book covering our home city of Sheffield. As with the Supertram book, the purpose of this publication is to cover the history of Manchester's first generation tramway, the background to the Metrolink project and its present day operations, tram fleet and potential future plans.

Compared to some other cities such as Sheffield, the history of Manchester's past and present day tram networks is much more complex: the city's first generation tramway was physically connected to, and had through running to and from, a number of neighbouring systems. Likewise, the present day Metrolink system is not simply a conventional tramway operating entirely within the Manchester city boundaries; rather it is an interurban network linking the city with nearby satellite towns within the Greater Manchester conurbation and including a mixture of on-street running, dedicated rights of way and former heavy rail lines. With Manchester having been the first major city in the UK to abandon its tramway after World War II and the first to reintroduce trams, it would seem fitting to make this the next book in the series especially with the latest extension to Trafford Park having opened in March 2020.

This book draws on material from a number of earlier Platform 5 publications, including our UK Metro & Light Rail Systems handbook; David Holt's Manchester Metrolink book published in 1992 to coincide with the opening of Phase 1; the various editions of our Light Rail Review series published in the late 1980s and 1990s; back copies of **entrain** and **Today's Railways UK** magazines; the Manchester Metrolink Handbook by John Senior and Colin Reeve and my father Ian Yearsley's book The Manchester Tramways, both of which were published by the Transport Publishing Company/Venture Publications; and official Greater Manchester PTE/Transport for Greater Manchester and Metrolink literature both from recent years and from the time that Metrolink was being planned.

I have made every effort to ensure that all information is correct at the time of going to press but cannot be held responsible for any errors or omissions. Nonetheless any corrections or suggestions for improvements for any future editions would be most gratefully received. Any comments on this publication should be sent by email to **updates@platform5.com** or by post to the Platform 5 address on the title page.

ACKNOWLEDGEMENTS

Thanks are given to all the individuals who have helped with the compilation of this book. We are particularly indebted to David Holt, the author of our original Metrolink book in 1992, along with Andrew Macfarlane, for their extensive knowledge of the history of the network and my father, Ian Yearsley, for his memories of Manchester's first generation tramway. Thanks are also due to Paul Abell, Dennis Gill and Steve Hyde for their assistance in sourcing archive material and photographs. If readers have any photos that they would like to be considered for our forthcoming tram system guidebooks (particularly any illustrating the early years) please do get in touch using the email **pictures@platform5.com**.

UPDATES

Any major developments with Metrolink, and the country's other light rail and tram systems, can be found in the magazine **Today's Railways UK**. This is available at all good newsagents or on post-free subscription. Please see the inside covers of this book for further details.

Alan Yearsley. March 2021.

Below: M5000 car 3086 departs from Moor Road with a Manchester Airport service on 8 November 2014, five days after the start of passenger services on the Airport line. *Robert Pritchard*

CHAPTER I:

MANCHESTER'S FIRST TRAMS:
72 YEARS OF SERVICE

Manchester's first generation tram network had its origins in the Manchester Suburban Tramways Company, established in 1877 to operate horse tram services in Manchester and Salford with the company's first tram service having operated on 17 May that year. In 1880 the company merged with horse bus operator the Manchester Carriage Company to form the Manchester Carriage and Tramways Company, which had developed a tram network of over 140 route miles by 1900.

In 1895 Manchester Corporation agreed to take over and modernise the city's tram network by the early 1900s, with the MC&TC's leases due to expire between 1898 and 1901. The Corporation sent delegations to view the methods of propulsion used elsewhere to decide which system would be best for Manchester. Systems examined for potential use were storage batteries, underground conduit, cable haulage (as used in Edinburgh at that time), steam (used for Leeds trams), gas (used in Lytham St Annes) and oil. There was even a delegation sent to examine the compressed air system in use on the Paris tramways. It was then decided to adopt a conventional overhead wire system common to most other tramways in Britain. Two of Manchester's neighbouring tramways introduced electric operation as early as 1899: the Oldham, Ashton and Hyde Tramway Company launched electric tram routes serving Ashton-under-Lyne, Denton, Hathershaw and Hyde on 12 June that year, followed by Bolton Corporation in December (with public service starting on 1 January 1900). Although Manchester itself only ever operated horse and electric trams, steam trams operated in Rochdale from 1883 until 1902 and gas trams in Trafford Park from 1897 until 1908 (which

were then replaced by trains hauled by steam tram engines). Both the Lytham and the Trafford Park gas trams were run by the British Gas Traction Company. There were also some trial runs in 1898 of a steam tram developed by French steam traction pioneer Léon Serpollet between Rusholme and Manchester city centre and in Trafford Park.

A new depot equipped for electric operation was built on Queens Road, Cheetham (part of which is now the Greater Manchester Museum of Transport). A second depot at Hyde Road, Ardwick was opened at the end of 1902, followed by a third depot at Princess Road, Moss Side in June 1909. The first overhead power lines were installed on the route between Cheetham Hill and Albert Square, which opened on 7 June 1901. By the end of that year further sections had opened between Cheetham Hill and Rochdale Road; Deansgate and Hightown; and High Street and Blackley/Moston Lane/Queens Park. On 1 June 1902 the tram tracks reached Piccadilly where the Corporation Tramway offices (now the Gardens Hotel) were located. 13 April 1903 saw the last horse trams in Manchester, and by the end of that year the city boasted tram routes from Piccadilly to Alexandra Park, Audenshaw, Denton, Hollinwood, Moss Side, Newton Heath, Old Trafford, Openshaw, St Peter's Square and Stretford. By this time the tram fleet had grown to just over 300 cars.

In 1899, two years before the start of electric tram operation, Manchester Corporation purchased six sample trams for evaluation purposes, numbered 101–106. Cars 101, 102, 104 and 105 were

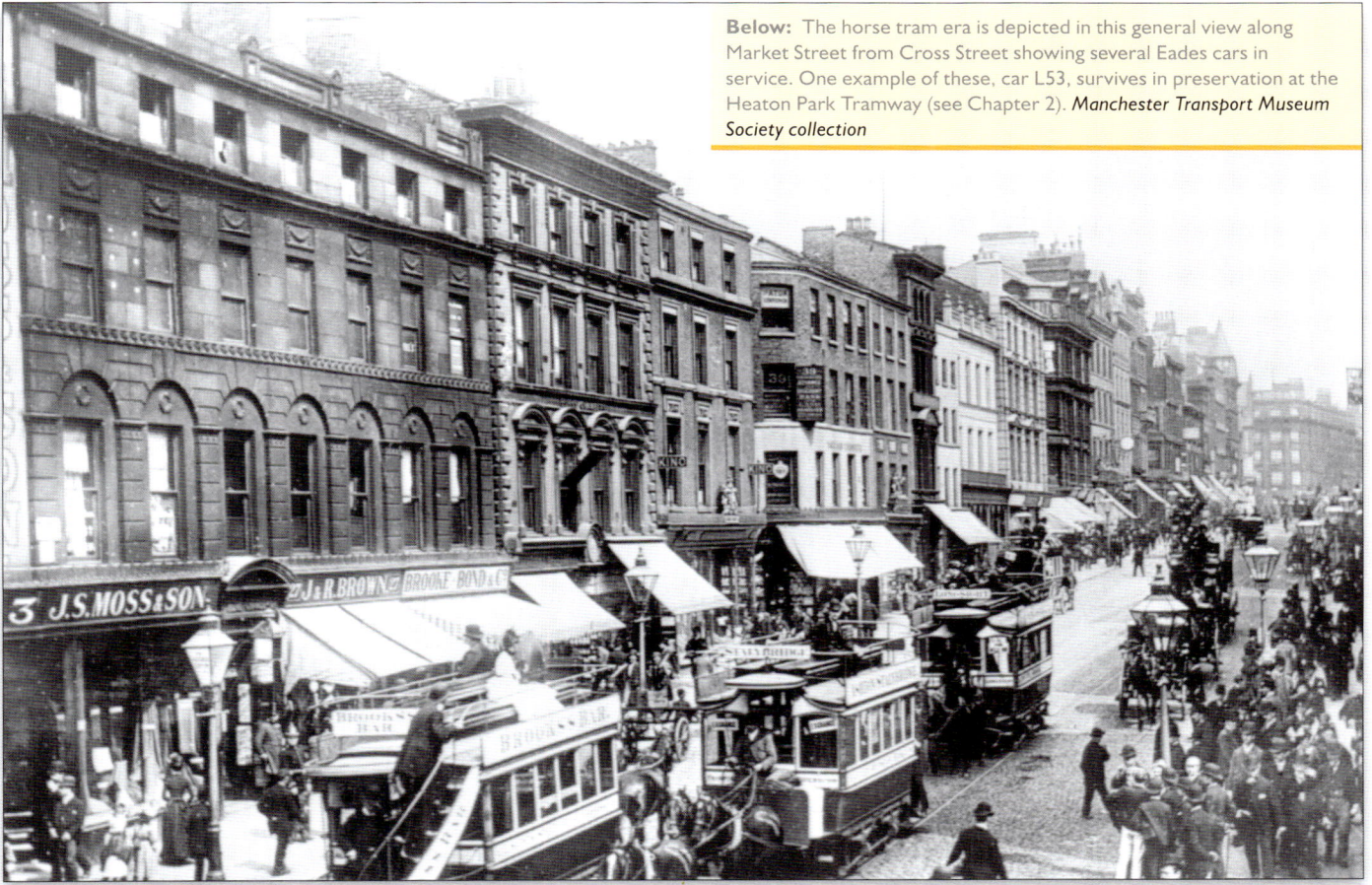

Below: The horse tram era is depicted in this general view along Market Street from Cross Street showing several Eades cars in service. One example of these, car L53, survives in preservation at the Heaton Park Tramway (see Chapter 2). *Manchester Transport Museum Society collection*

Above: The procession of trams in Albert Square for the official opening of the electrified system on 7 June 1901. *Manchester Transport Museum Society collection*

Above: A line-up of trams including open balcony car 749 (left) and fully enclosed car 947 (right) on Mount Road siding wait to pick up crowds of visitors to the adjacent Belle Vue Gardens, the site of a popular zoo and amusement park between the 1830s and the 1970s. *Manchester Transport Museum Society collection*

Above: A busy scene in Piccadilly with a line of cars waiting to leave George Street. The large building in the background is the current Britannia Hotel on Portland Street. *Manchester Transport Museum Society collection*

Above: Deansgate with Salford car 312 heading for Whitefield and a Manchester car behind it. In the background can be seen the tower of Manchester Cathedral. *Manchester Transport Museum Society collection*

open-top four-wheelers, with 101 being built by Manchester Carriage & Tramways Company itself, 102 by Hurst Nelson of Motherwell, Scotland, 104 by the Ashbury Carriage and Iron Company in Manchester and 105 by G.F. Milnes & Co based in Birkenhead and Hadley, Shropshire. 103 was an open-top bogie car built by Brush and 106 was a single-deck four-wheeler built by G.F. Milnes. The first

production series cars started to enter service in 1901, consisting of open-top four-wheelers numbered 107–187 (built by Brush) and 277–436 (Milnes) and open-top bogie cars 437–486 (Brush). Over the years several more batches followed. Initially a mixture of four-wheelers and bogie cars were purchased, but almost all orders from 1913 onwards were for bogie vehicles. The first top-covered double-deckers with

Above: This photo taken during Civic Week 1926 shows skeleton tram 992, which ran around parts of the system to show the construction methods of the contemporary double deck bogie car. *Manchester Transport Museum Society collection*

open balconies at either end of the top deck were delivered in 1904 and the first fully enclosed double-deckers in 1920. Six years later, in 1926, the highest-numbered cars, 1004–1053, entered service. These were fully enclosed double-deck bogie cars built by English Electric. However, these were not the last of Manchester's first generation trams to be built, because the Corporation purchased a total of 148 new fully enclosed double-deckers between 1924 and 1932 to replace withdrawn trams of early 1900s vintage, in each case taking the fleet number of the car that they replaced. In all cases the original vehicle was a four-wheeler, except for 220 which was a bogie car.

The last 49 of these, delivered between 1930 and 1932, were known as Pullman cars because of their stylish design. They were also nicknamed Pilcher cars after the then General Manager Stuart Pilcher (see below). Rather oddly all of these were four-wheelers, the last such vehicles from the original batch having been delivered in 1912. A notable feature of all Manchester Corporation trams was that the designs were modular, with standard window spacing both for four-wheelers and for bogie cars. This enabled four-wheel cars to be lengthened to bogie vehicles from 1920 onwards.

In the early years of the 20th century Manchester's neighbouring boroughs such as Ashton-under-Lyne, Bury, Oldham, Salford and Stockport operated their own tram networks and began their own modernisation programmes. Salford Corporation started running trams down Deansgate, Manchester, in 1903. By the 1910s and 1920s the tram networks across what is now the Greater Manchester conurbation were rapidly expanding and several more through routes between networks were being established, with many such services being operated jointly by the respective operators. For example, Stockport trams ran into Manchester on the routes to Hazel Grove and Hyde. This meant that it was possible to travel by tram all over Greater Manchester and into the nearby towns in Cheshire and Lancashire, many of which had their own independent tram networks. From

1921 Manchester trams ran through to Ashton-under-Lyne via Guide Bridge. This was made possible by the acquisition by Manchester Corporation, jointly with Ashton Corporation and Stalybridge Joint Board, of the the Oldham, Ashton and Hyde Tramway. Then in 1923 Chadderton, Middleton and Rochdale councils jointly took over the Middleton Electric Traction Company, with Middleton granting Manchester Corporation a lease to run on their former tracks. This enabled Manchester trams to run all the way through to Rochdale. Manchester's last new tram route opened in 1928, connecting the city to Middleton's line to Hopwood, Heywood, the boundary with the Bury Corporation network. The same year also saw the opening of Manchester's fourth and last new tram depot at Birchfields Road, and by this time the Manchester Corporation tramway network itself had reached its maximum size of 123 miles of route (292 track miles) with a fleet of 953 trams, making it the third largest system in Britain after London and Birmingham.

DECLINE AND CLOSURE

It is often said that the heyday of tramways in the UK occurred between the 1900s and the 1920s, with the decline starting in the 1930s, and Manchester was no exception. In 1929, following the unexpected death of General Manager Henry Mattinson, Stuart Pilcher was appointed as his successor. Unlike Mattinson who was pro-tram, Pilcher believed that the tram was becoming an outdated form of transport and should be phased out and replaced by buses and trolleybuses. To reflect this change of policy, Manchester Corporation Tramways was renamed Manchester Corporation Transport. In 1930 circular route 53 from Cheetham Hill to Stretford Road became the first tram route to close and be replaced by buses, and plans for a tramway extension to serve the rapidly expanding council housing estate of Wythenshawe in south Manchester were abandoned. Also never realised were proposals for underground tram routes across the

Above: California car 838 on route 53 to Brooks's Bar encounters the replacement motor buses on Chorlton Road on the first day that buses operated for a full day in 1930. *Manchester Transport Museum Society collection*

city centre that were mooted on at least two separate occasions in 1908 and 1914.

In 1937 the Corporation decided to abandon its tram network completely and in 1938 the first trolleybuses ran on former tram route 28 from Piccadilly to Ashton. It is likely that World War II delayed the closure of the city's tram network, and in 1942 Manchester Corporation reinstated services 13 (Albert Square–Chorlton–Southern Cemetery) and 23 (Hollinwood–Chorlton via Brooks's Bar) and Salford reinstated part of service 76 (Deansgate–Weaste

Left: 1007, Manchester's official last tram, enters Birchfields Road Depot at the end of the last trip on 9 January 1949. *Manchester Transport Museum Society collection*

were removed during the night of 9–10 January, meaning that the rush hour trams on the morning of Monday 10th observed the bus stops instead. Most of these cars ran only as far as Albert Road, Levenshulme, but one of them ran all the way to the Manchester city boundary at Lloyd Road. All carried route 37 number stencils.

The last tram route in Greater Manchester to be abandoned was that of Stockport Corporation serving Reddish, which closed in 1951. Trolleybuses continued to operate in Manchester and Ashton-under-Lyne until 1966, after which the city relied entirely on buses and suburban trains to meet its transport needs for the next 26 years.

via Cross Lane). From that date Manchester route 37 was extended from the old Southern Cemetery crossover at Maitland Avenue to a new one at Princess Road. New track was laid in Albert Square to enable the reinstated service 13 to make better use of the square as a terminus.

Nonetheless, Manchester became the UK's first major city to abandon its tramways after the war with the last trams running in normal service on 9 January 1949. Cars 113, 976 and 978 took part in the farewell procession, with car 1007 being the official decorated last car. In the event this was not quite the end because a shortage of buses the next day led to an unexpected use of trams in their place in the morning peak period between Piccadilly and Levenshulme. Tram stop signs

Right: The crews associated with the last tram, presumably at Birchfields Road Depot. *Manchester Transport Museum Society collection*

CHAPTER 2:

MANCHESTER'S TRAMWAY
HERITAGE

Above: "California" car 765 on display in Albert Square, Manchester on 5 November 2014 in connection with the centenary of the outbreak of World War I. *David Holt*

Given the determination of Manchester Corporation to rid Manchester of its trams as soon as possible after World War II (including the almost immediate scrapping of most of the tram fleet), and that the city's first generation tramway closed before there was much interest in preserving tramcars, it is fortunate that three of Manchester's old trams did nonetheless survive (although unfortunately none of the city's trams built later than 1914, including the Pilcher cars, were saved for preservation). All three of these are based at the Heaton Park Tramway.

Reputedly the largest public park in Europe, Heaton Park was sold to Manchester Corporation in 1902 by the Earl of Wilton. In 1903 the city's tram network was extended into the park from the Middleton Road entrance for special excursion trams from various parts of the city, mainly for Sunday School parties. In 1934 the Heaton Park tramway siding closed and was covered with tarmac. During the 1970s the Manchester Transport Museum Society approached Manchester City Council with a view to reopening the former tramway siding, and following negotiations work started on unearthing the existing tram rails beneath the tarmac. The tramway was officially opened by the Lord Mayor of Manchester at Easter 1980. Trams initially operated only between the Middleton

Road entrance and the former tram shelter, which was converted into a depot. Over the years the line has been extended three times and is now about 1 km in length having reached its current terminus adjacent to the Pavilion Cafe by the boating lake in 2011. Trams normally operate on Sundays and bank holidays from late March until mid-November and on Saturdays in July and August. Full details of the tramway, its history, tram fleet, extension plans and special events can be found on the website at *www.hptramway.co.uk*.

765

The only first generation Manchester electric tram currently in working order is single deck car 765, one of a third series of combination cars (referred to as such because of their mixture of open and enclosed seating areas) built for routes with low bridges. These vehicles had a central saloon and open platforms at each end with seats on which smoking was permitted. They were based on an American design and were thus sometimes referred to as "California" cars.

Above: Horse tram L53 at Heaton Park on 28 March 2010. *David Holt*

Car 765 was built by the United Electric Car Company in Preston in 1914, with the bogies being manufactured by G.F. Milnes. Final assembly of the vehicle took place at Manchester Corporation's Hyde Road Car Works. It operated on circular route 53 from Cheetham Hill to Stretford Road via Belle Vue throughout its working life until closure of the route in 1930 after which most of the combination cars were sold or scrapped. Only six of them remained by 1931 and were now confined to works duties. After car 847 was scrapped in 1948 the only remaining example of this type was car 765, whose body was used as a hen house at Pioneer Farm at Blackmoorfoot near Huddersfield for several years. In 1960, members of the then Manchester Transport Historical Collection (later to become the Manchester Transport Museum Society (MTMS)) set about preserving the tram. After being stored at the National Tramway Museum at Crich for a time, car 765 was moved to Birchfields Road depot for restoration and ran at Crich for a short period in the late 1970s. It then returned to Manchester in 1979, since when it has operated on the Heaton Park Tramway. Because its original bogies did not survive when the car was withdrawn from service, its present day bogies are regauged ones from the Hill of Howth Tramway in Ireland.

Although based at Heaton Park, 765 visited Blackpool in 1985 for the town's tramway cententary and in 2010 for the 125th anniversary celebrations, and also made an appearance at the Great North Steam Fair at Beamish Museum in County Durham in April 2011 and again in 2019.

HORSE TRAM L53

The other operational first generation Manchester tram is Manchester Carriage & Tramways Company horse car L53, built in the 1880s. Its remains were discovered near Glossop, Derbyshire in 1970 where it had been used as a hairdresser's and a fish and chip shop among many other uses. Unlike most other tram bodies that were used as sheds, greenhouses etc, L53 was still on its wheels. Restoration of L53 took place in several different locations, with most of the side frames being made by an MTMS member as part of an A-level woodwork project. In 1998 the vehicle arrived at Heaton Park, and further restoration work was then carried out on site over the next ten years. The tram was relaunched into service at a commemorative event at Heaton Park on 27 March 2008. L53 operated at Beamish in May 2009, and in the same year it won the "Best Tram" award in the Heritage Railway Association's annual Carriage and Wagon Competition.

L53 is one of over 500 open-top horse trams designed by John Eades in 1877 and ran in service until 1903. John Eades was Manager of the Manchester Carriage Company's coachbuilding works at Ford Lane, Pendleton, Salford from 1867 until 1903. A unique feature of the Eades horse trams was the ability of the horse to rotate the body on its underframe. This saved time when turning round at each end of the route, as most other horse cars required the horse to be uncoupled

Right: Manchester car 173 on display outside Victoria station on 21 October 2019 in connection with the station's 175th anniversary. An unidentified Metrolink M5000 tram can be seen in the background. *David Holt*

Left: Stockport car 5 at Heaton Park on 17 November 2013, with Mayor of Stockport Cllr Chris Murphy driving (Vernon Park was in his ward).

Below: Hull car 96 at Heaton Park during a visit by the Leeds Transport Historical Society on 11 July 2017, with the late Derek Shepherd of the Manchester Transport Museum Society driving. *Paul Abell (2)*

on arrival at the terminus. L53 also had only one staircase unlike most other double deck trams, thus reducing the weight of the vehicle and allowing a higher seating capacity. Although nominally still part of the Heaton Park Tramway fleet, since 2010 L53 has resided at Bury Transport Museum.

The chassis of Manchester horse tram W24, also built by Eades, survives at Crich and is on display in unrestored condition in the exhibition hall at the museum. This tram ran in service from 1880 until 1903. Its history from that date is unknown until 1961 when the vehicle arrived at Crich. Its body was in such a dilapidated condition that it was soon scrapped. An unrelated tram body was acquired and stored off-site in 2000.

TABLE 1: THE HEATON PARK TRAM FLEET

Home town/city and fleet number	Type	Built by	Status	Notes
Stockport 5	Four-wheel open topper	Dick Kerr, 1901	Operational	
Rawtenstall 23	Four-wheel single decker	UEC, 1912	Awaiting restoration	
Oldham 43	Four-wheel single decker	ERTCW, 1902	Awaiting restoration	Converted from open topper 1933.
Manchester L53	Four-wheel open top horse tram	MCTC, 1880	Operational	Currently at Bury Transport Museum.
Hull 96	Four-wheel single decker	Hurst Nelson, 1901	Operational	Sold to Leeds 1942–45, ran as stores car 6 until 1959.
Manchester 173	Four-wheel open topper	Brush, 1901	Cosmetically restored, awaiting overhaul	
Blackpool & Fleetwood 619	Bogie single decker	Mode Wheel works, Salford, 1987	Awaiting repairs	
Blackpool 623	Bogie single decker	Brush, 1937	Operational	
Blackpool 680	Bogie single decker	English Electric, 1935	Operational	On loan to Blackpool Tramway.
Blackpool 702	Bogie double decker	English Electric, 1934	Awaiting restoration	Stored in Sunderland.
Blackpool 708	Bogie double decker	English Electric, 1934	Awaiting restoration	Stored at Rigby Road Depot, Blackpool.
Blackpool 752	Four-wheel railgrinder	Blackpool Corporation, 1928	Awaiting restoration	
Manchester 765	Bogie single decker	MCTD, 1914	Operational	
Manchester Metrolink 1007	Bogie articulated single decker	Firema, Italy, 1991	Stored	Stored at Metrolink's Trafford Depot.

event in December 2020 after a seven-year overhaul. Oldham 43 was built as an open topper but converted to a single decker in 1933 for the Oldham–Middleton route. Restoration of this vehicle is a long-term project.

OTHERS

Heaton Park is also the home of a number of trams from other networks across the north of England. The only other two trams currently in working order and in regular use are Blackpool Brush Railcoach 623 and Hull 96. The table lists all trams in the Heaton Park fleet (some of which reside elsewhere at the time of writing). Probably the vehicle with the most remarkable history is Blackpool &

173

Manchester Corporation car 173 is a four-wheel double decker built by Brush in 1901. Originally an open-top car, it later received a top cover but retained open balconies. 173 remained in service until 1931, after which it was used as a garden shed for many years. Restoration of the tram was carried out at a number of different locations including the Manpower Services Workshop at Horwich Loco Works. After a period of being based at the Greater Manchester Museum of Transport at Boyle Street, Cheetham Hill, the tram arrived at Heaton Park in December 2013. It has now been cosmetically restored to its original condition as an open-topper but will require further work to return it to working order. The tram normally lives in the depot at Heaton Park but is occasionally put on static display for special events. In October 2019 it was displayed at Manchester Victoria station as part of a special event to celebrate the station's 175th anniversary.

Fleetwood 619, which was originally built as Railcoach 282 in 1935. Renumbered 619 in 1968, it was converted to a one-man-operated (OMO) car in 1973 when it returned to service and ran in this form until 1987. It was then rebuilt as a replica Blackpool & Fleetwood vanguard tram, a type of "toastrack" car with open sides. However, a number of modifications were made compared to the original design such as a central aisle instead of solid transverse seating. This enabled passengers to board and alight via the rear platform as on a conventional single or double decker. Open sides were retained but with railings fitted at waist height for safety. 619 ran in Blackpool until 2004 and again during the 2008 illuminations season, and has resided at Heaton Park since 2010.

STOCKPORT 5 AND OLDHAM 43

Heaton Park also has two other trams from first generation tramways in Greater Manchester: Stockport 5 and Oldham 43. Stockport 5 is an open topper built in 1901, which was fully restored to operational condition in time for the tramway's "Lightopia" Christmas lights

CHAPTER 3:

OVERGROUND, UNDERGROUND

Above: An artist's impression of a Duorail car as proposed by the LRTA under the MARTIC scheme, reproduced from a mid-1960s MARTIC brochure. Note that this vehicle has a short centre section unlike the Metrolink trams that eventually materialised.

In common with many other cities throughout the developed world, Manchester saw a significant increase in motor car traffic during the 1950s and 1960s following the closure of its tram network. At this time the planning authorities regarded the trend away from mass transport on buses, trams and trains towards individual car transport as progress, and accordingly the 1962 South-East Lancashire North-East Cheshire (SELNEC) Highway Plan, devised by members of the SELNEC Area Highway Engineering Committee, envisaged a comprehensive programme of road building and widening of existing roads, with segregated pedestrian walkways being provided to keep pedestrians separate from road vehicles wherever possible.

However, despite the increasing use of the private car for commuting and non-work journeys, there remained a problem of poor connectivity between the rail networks to the north and the south of the city (served from Victoria and Piccadilly stations respectively) with many passengers needing to transfer between the two stations by bus, taxi or on foot. Both stations are located at the edge of the city centre, some distance from many shops, workplaces, eating and drinking establishments, entertainment venues, hotels and visitor attractions. This lack of connectivity between the two rail networks was a legacy of the competition between railway companies prevalent in the Victorian era.

Because of this, in the mid-1960s local representatives of the Light Railway Transport League (now the Light Rail Transit Association) devised a light rail scheme known as "MARTIC" (Manchester Area Rapid Transit Investigation Committee) in response to the lack of any light rail-based options in the SELNEC Transportation Study set up in 1965. The MARTIC scheme had many similarities with the present day Metrolink network, consisting of a north–south route from Bury to Wythenshawe and Manchester Airport (then known as Ringway Airport) including an on-street city centre section. This scheme was promoted as "Duorail" to distinguish it from the monorail concept involving trains running on just one rail (see below).

Then in 1966 a study group from Manchester City Council's transport, highways and planning committees examined a variety of rapid transit systems across the world and concluded that the French Safege monorail system, on which trains run suspended by the roof from the rail on which they run, would offer the best solution. They then proposed to build a 16-mile monorail from Langley, a district of Middleton to the north of the city to Manchester Airport in the south via Victoria station, the former Central station, Rusholme, West Didsbury, Northenden and Wythenshawe, following a similar route to that envisaged in the MARTIC plan. Trains would have run at up to 50 mph, with 19 stations at half-mile intervals. The route was planned to follow existing roads to the north and south of the city and serve the main business and entertainment areas in the city centre. Two different routes between Victoria and Central stations were considered, one running above ground, the other in tunnel.

This was followed in 1967 by a Manchester Rapid Transit Study compiled jointly by the Ministry of Transport, Manchester City Council and British Rail, which also considered various types of rapid transit system, including the Safege monorail, the Alweg monorail (on which vehicles run on top of a single rail rather than underneath), and the Westinghouse System consisting of rubber-tyred vehicles running on parallel concrete rails with a guide rail betweeen the two running rails.

PICC-VIC TUNNEL

In the event, the consultants concluded that a conventional heavy rail-based solution would offer best value for money. At this time it was envisaged that such a system would follow a route based broadly on that originally proposed for a monorail. The scheme was

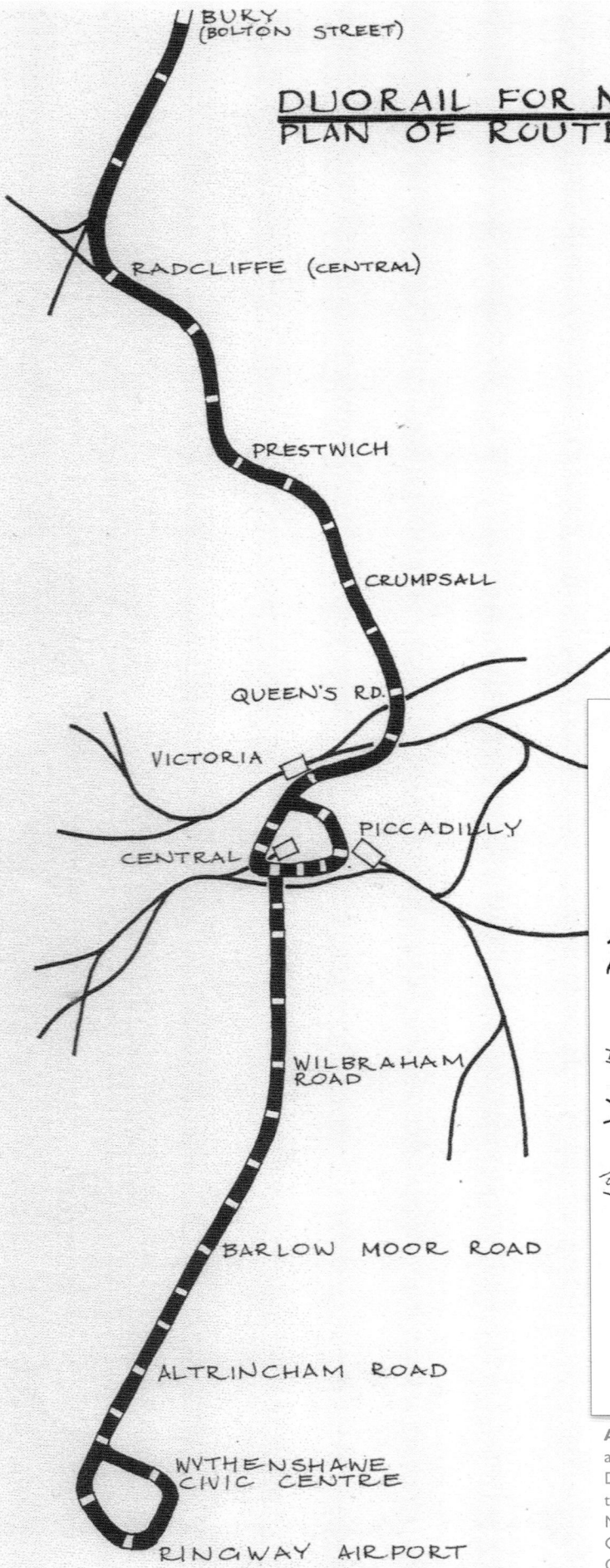

DUORAIL FOR MANCHESTER
PLAN OF ROUTE

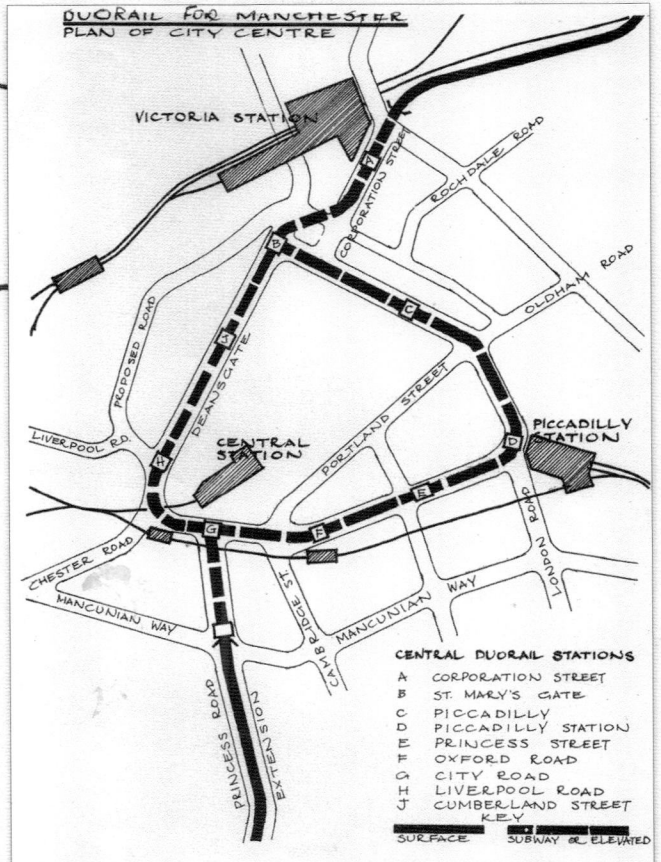

BURY
(BOLTON STREET)

RADCLIFFE (CENTRAL)

PRESTWICH

CRUMPSALL

QUEEN'S RD.

VICTORIA

CENTRAL

PICCADILLY

WILBRAHAM ROAD

BARLOW MOOR ROAD

ALTRINCHAM ROAD

WYTHENSHAWE CIVIC CENTRE

RINGWAY AIRPORT

DUORAIL FOR MANCHESTER
PLAN OF CITY CENTRE

VICTORIA STATION

CENTRAL STATION

PICCADILLY STATION

CORPORATION STREET
ROCHDALE ROAD
OLDHAM ROAD
PROPOSED ROAD
DEANSGATE
LIVERPOOL RD.
PORTLAND STREET
LONDON RD
CHESTER ROAD
MANCUNIAN WAY
CAMBRIDGE ST
MANCUNIAN WAY
PRINCESS ROAD
EXTENSION

CENTRAL DUORAIL STATIONS

A	CORPORATION STREET
B	ST. MARY'S GATE
C	PICCADILLY
D	PICCADILLY STATION
E	PRINCESS STREET
F	OXFORD ROAD
G	CITY ROAD
H	LIVERPOOL ROAD
J	CUMBERLAND STREET

KEY
SURFACE   SUBWAY or ELEVATED

Above: These two maps, taken from the same MARTIC brochure as the image on the facing page, show the proposed route of the Duorail scheme. In many ways the route closely resembled that of the present day Metrolink lines to Bury and Manchester Airport. Note that the present day Manchester Central (formerly G-Mex) Convention Centre was then still in existence as Manchester Central station.

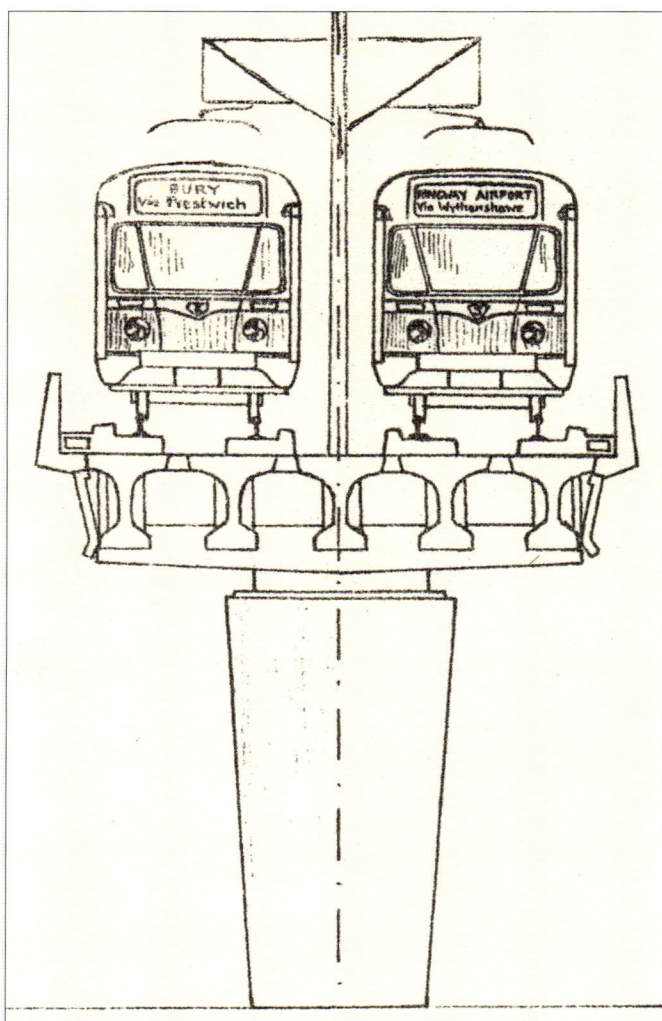

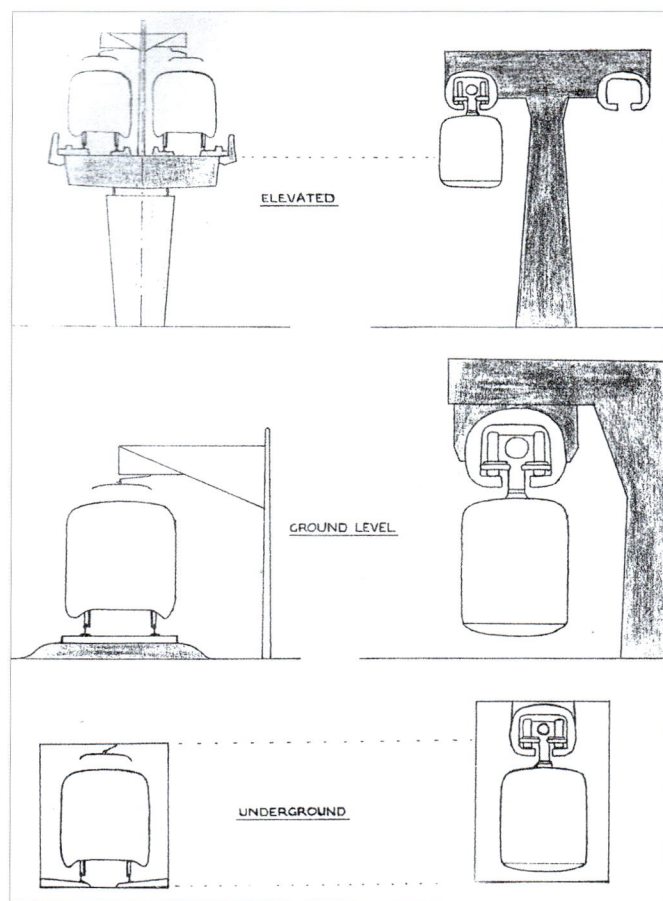

Above & above left: Artist's impressions of a Duorail elevated section (**left**) and comparisons of Duorail and monorail (**above**) showing elevated, ground level and tunnel sections, taken from the same MARTIC brochure as the image and maps on the previous two pages.

not pursued any further in that form, but by the end of the 1960s SELNEC Passenger Transport Executive (which later became Greater Manchester PTE) had unveiled a project to build a tunnel linking the two main stations to enable through running between destinations in the north and south, appropriately known as the Picc-Vic Tunnel. This was not the first project to connect the two stations via a railway tunnel, with such a scheme having first been mooted as early as 1839 in anticipation of the opening of Victoria and London Road (now Piccadilly) stations. The idea of a circular tunnel linking the main stations was also floated in the early 1900s and in 1912, and more ambitious schemes for underground rail networks serving the city centre and surrounding areas were proposed in the 1920s but never came to fruition.

The new 2.75 mile Picc-Vic line would have diverged from the main line into Piccadilly at Ardwick Junction, just south of Piccadilly station, and joined the Bury line at Queens Road Junction about 0.75 mile north of Victoria. Just over 2 miles (3 km) of the route would have been in tunnel. The proposed route would have skirted round the southern edge of the city centre as far as Albert Square before heading north towards Victoria. There were to be five new stations at Piccadilly Low Level, Princess Street, Albert Square/St Peter's Square, Market Street and Victoria Low Level. Alternative names considered for the three intermediate stations were Whitworth, Central and Royal Exchange respectively. Each route was to have operated at a 10-minute frequency, giving a 2.5 minute metroval service on the central section. Travelators (moving walkways) were planned to link Piccadilly station with Piccadilly Gardens and Albert Square/St Peter's Square station with the existing Oxford Road station. Four routes were planned to serve the tunnel section: Hazel Grove and Macclesfield to Victoria (with a potential extension to Oldham and Rochdale at a later date), Alderley Edge–Bury via Stockport, and Wilmslow–Bolton via Styal

using the Bury–Bolton line, which had closed to passengers in 1970. BR planned to order new trains for the Picc-Vic services, which were to be known as Class 316. These would have been based on the Class 313/314/315/507/508 family of units, which were derived from the Southern Region "Pep" prototypes. The 316s would have replaced the existing 25 kV AC slam-door Class 304 EMUs used on south Manchester services and the 504s on the Bury line, which would have been converted from 1200 V DC side contact third rail to 25 kV AC overhead. Also included in the plans was provision for fill-in light rail services. In fact, one local Light Rail Transit Association member suggested conversion of the Oldham Loop to light rail with on-street links into and across Manchester city centre and an on-street extension into Rochdale town centre, an idea that would eventually come to fruition with the present day Metrolink system.

The Picc-Vic scheme received parliamentary powers in 1972, and SELNEC PTE made an infrastructure grant application to the Government to fund the project. However, the economic downturn of the early 1970s led the then Chancellor of the Exchequer Anthony Barber to announce a £500 million reduction in public spending, and Transport Minister John Peyton rejected SELNEC's grant application. The scheme was eventually abandoned in 1977 on cost grounds, although the new bus/rail interchanges at Altrincham and Bury and electrification of the Hazel Grove line, which had been envisaged as part of the Picc-Vic project, did still go ahead. Also built in preparation for the scheme was a subterranean void beneath the Arndale Centre to accommodate the proposed Market Street/Royal Exchange station.

A potential obstacle to the construction of the line was Guardian, an underground telecommunications complex to the south of Piccadilly Gardens, which had been designated as a Cold War nuclear bunker. SELNEC planners were aware of this structure but were unable to publicise its existence, as it was protected by a "D Notice"

The Picc-Vic Project
GMC ᛗ

Left: The front cover of GMPTE's 1975 promotional brochure for the Picc-Vic scheme featured an artist's impression of a Class 316 EMU on a Bolton-bound service calling at the proposed Market Street/ Royal Exchange station.

Below: This diagrammatic map of the proposed routes, taken from the same brochure, shows how the scheme would have funnelled commuter services into the city centre tunnel section.

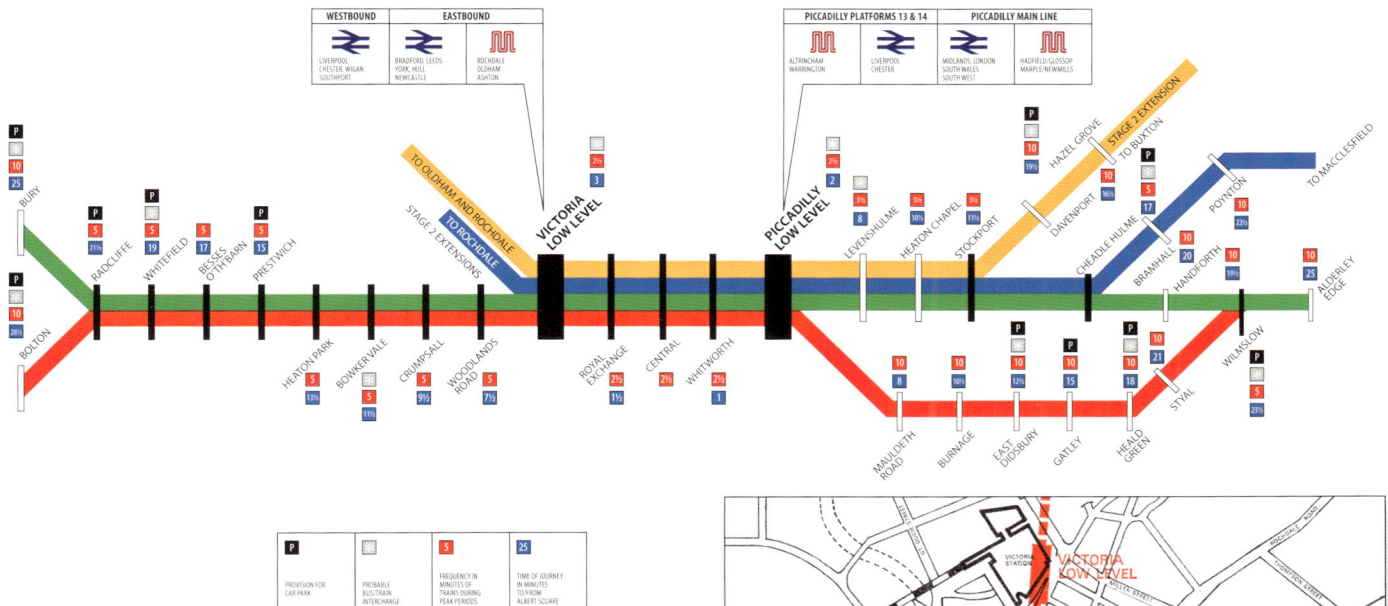

in the 1950s to prevent the media from writing about it and even when the Picc-Vic project was being evaluated SELNEC was required to sign the Official Secrets Act, which meant that its exact location could not be disclosed.

A partial solution to the lack of connectivity between the north and south Manchester rail networks eventually materialised in 1988 in the form of the Castlefield Curve (a.k.a. Windsor Link), which runs from Windsor Bridge Junction, near Salford Crescent station, via Ordsall Lane Junction to Castlefield Junction. This has enabled a number of east–west services that previously used Victoria station to be rerouted to Piccadilly, and was complemented by the new Ordsall Chord in 2017 giving a direct railway link between Piccadilly and Victoria stations for the first time.

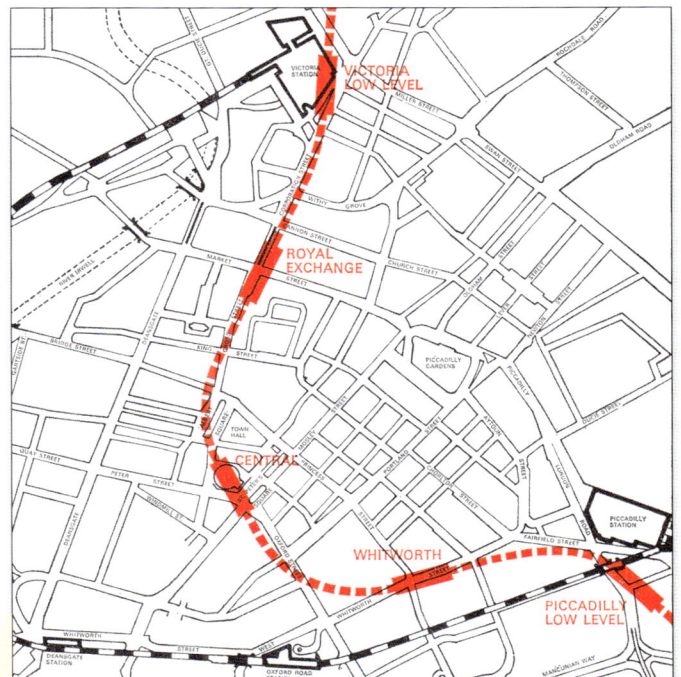

Right: This map, also taken from the same brochure, shows the proposed route of the Picc-Vic Tunnel through Manchester city centre.

CHAPTER 4:

TRAMWAY
REVIVAL

Following the demise of the Picc-Vic scheme, in 1982 British Rail, Greater Manchester Council and Greater Manchester PTE set up a joint Rail Study Group to examine a number of options to improve connectivity to, from and within Manchester. In April 1983 the group published an initial report entitled "A Rail Strategy for Greater Manchester: The Options – First Report of the Rail Study Group". This document considered four possible ways forward:

- A conventional heavy rail-based solution, including east–west and north–south city centre tunnels and a possible reinstatement of the "Midland Curve" from Philips Park No. 1 Junction into Piccadilly via Ardwick Junction to allow Calder Valley services to use Piccadilly instead of Victoria station.
- An on-street light rail system in the city centre.
- A light rail system operating in tunnels under the city centre.
- A guided and/or conventional busway.

In 1984, the Rail Study Group presented abstract proposals for a 62 mile (100 km) light rail network to the Government for funding. This consisted of three routes, all of which would involve taking over existing or disused heavy rail lines linked by on-street tracks through the city centre:

- Line A: Altrincham–Hadfield/Glossop
- Line B: Bury–Rose Hill Marple
- Line C: Rochdale–East Didsbury via the Oldham Loop and the former Manchester South District Line.

In November 1984 GMPTE deposited the UK's first modern Parliamentary Bill promoting on-street light rail operation, entitled "The Greater Manchester (Light Rapid Transit System) Bill". This was followed by a second Bill in May 1986 covering work on the Bury and Altrincham lines. Both Bills received Royal Assent on 9 February 1988. Meanwhile in July 1985 GMPTE applied for Section 56 funding towards the £42.5 million projected capital cost of Phase 1 of the scheme, covering the Bury and Altrincham lines. As well as the original tram fleet and conversion of the existing heavy rail lines, the funds applied for were also to go towards the cost of:

- Construction of new track and overhead wires between Cornbrook Junction and G-Mex (now Deansgate-Castlefield)

Above: An artist's impression of a Docklands Light Railway P86 car in Greater Manchester PTE orange. DLR car 11 was fitted with a pantograph for its visit to the demonstration line on the former Fallowfield Loop but remained in the DLR's then corporate blue/red livery as shown in the other photo. *GMPTE*

ROCHDALE

BURY

BURY SIDINGS

RADCLIFFE

WHITEFIELD

BESSES O'TH' BARN

PRESTWICH

HEATON PARK

OLDHAM

BOWKER VALE

CRUMPSALL

WOODLANDS ROAD

QUEENS ROAD

VICTORIA

PICCADILLY GARDENS

G. MEX

PICCADILLY

Salford Quays

DUMPLINGTON

TRAFFORD BAR

Trafford Park

OLD TRAFFORD

STRETFORD

Station

DANE ROAD

Phase 1

SALE

BROOKLANDS

Future Phase

TIMPERLEY

NAVIGATION ROAD

ALTRINCHAM

Light Rapid Transit

in Greater Manchester...

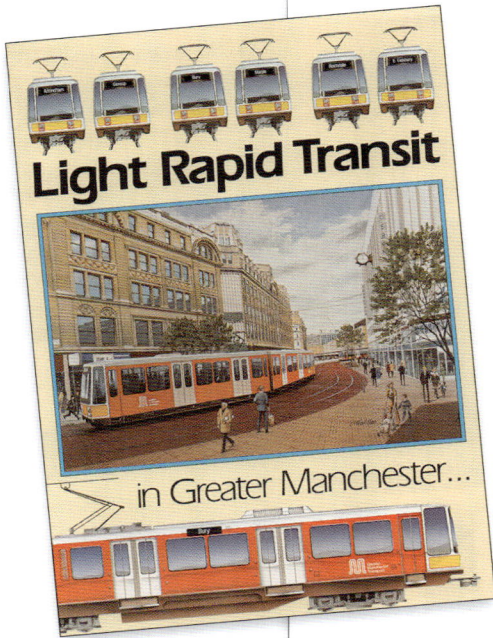

Above & below: Two 1980s GMPTE publicity brochures promoting the light rail scheme.

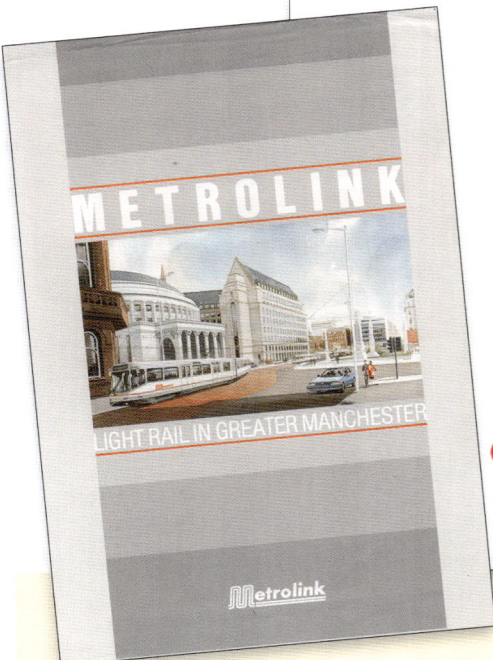

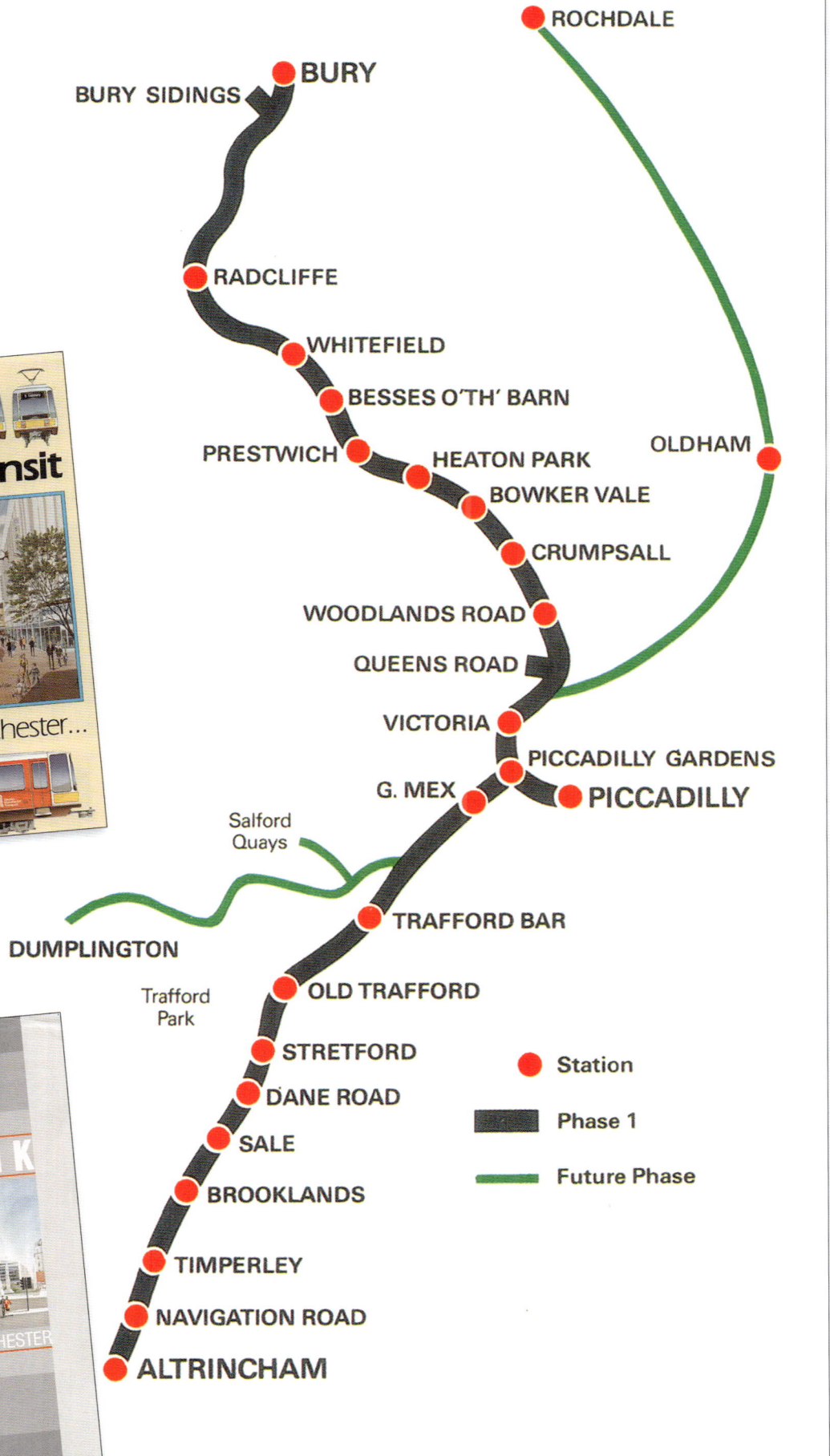

METROLINK

LIGHT RAIL IN GREATER MANCHESTER

Metrolink

Above: A map of Metrolink Phase 1, showing the then projected extensions to Oldham, Rochdale, Salford Quays (since extended to Eccles) and Dumplington (now the Trafford Park line) as shown in Platform 5 Publishing's 1992 Metrolink book by David Holt.

- Construction of an underpass at Cornbrook to take the Metrolink line beneath the Manchester–Warrington–Liverpool line
- Renovation of structures, especially the Cornbrook–G-Mex viaduct, which had carried the Cheshire Lines Committee tracks out of Manchester Central station until it closed in 1969
- Preliminary diversion of utilities in the city centre
- Enabling works off the line of route in the city centre such as traffic management and minor highway schemes
- Modifications to the computerised traffic control system
- Construction of new track, overhead wires and seven tram stops in the city centre
- A 750 V DC power supply covering the entire length of the Phase 1 routes
- Communications, signalling, fare collection and surveillance equipment
- An operational centre and maintenance depot.

GMPTE first adopted the name Metrolink in 1987 after securing funding and powers for Phase 1 of the network. At about this time the PTE proposed an extension to Rochdale town centre and a new line to Salford Quays and Dumplington.

PROJECT LIGHT RAIL

Between 1983 and 1985 the Rail Study Group undertook study visits to light rail networks in the Netherlands, Germany, Switzerland, France, the USA and Canada to find out what could be learned from tramways in those countries when designing such a system for the UK. This was followed by a further visit to France, the Netherlands and Switzerland in 1989. Meanwhile in 1986 British Rail Engineering (BREL) identified a light rail vehicle designed by the Urban Transportation Development Corporation in Canada for the Santa Clara Valley Transportation Authority in California, USA, as an example of the type of LRV that could be used on UK light rail systems. BREL had planned to ship a LRV of this type to the UK to take part in a live demonstration, but negotiations for the loan of this vehicle fell through. Instead Docklands Light Railway P86 car 11 was hired and used for a series of public demonstration runs in

March 1987, billed to the press as "Project Light Rail", on a 1 mile (1.6 km) stretch of track on the site of the former Hyde Road station on the Fallowfield Loop line, a now lifted freight line that once ran from Gorton and Fairfield to Manchester Central via Fallowfield (part of its former alignment now being used by the present day Metrolink line to East Didsbury and Manchester Airport). A temporary single-platform station was built at Debdale Park, and the DLR car was temporarily fitted with a pantograph and driven manually rather than operating driverless. New 750 V DC overhead power lines were also installed using masts designed by Balfour Beatty for the Tuen Mun light rail system in Hong Kong. This event was billed to the press as "Project Light Rail" and was a joint venture between GMPTE, British Rail, BREL, GEC Transportation Projects, Balfour Beatty, and Fairclough Civil Engineering. On 10 March, the first day of the event, Minister of State for Transport David Mitchell ceremonially flagged off the DLR car from the temporary platform. An exhibition on GMPTE's plans for Metrolink was also held adjacent to Debdale Park station. After the event the temporary platform was reused at the then new Hag Fold station between Daisy Hill and Atherton on the Manchester Victoria–Wigan Wallgate line.

MOCK-UP TRAMS

Then in 1988 a full-size mock-up of the proposed Metrolink tram design was constructed at Birchfields Road bus garage, originally opened as a tram depot in 1928. A mock-up profiled platform was also provided to trial the proposed strategy for achieving step-free access to two doorways while minimising physical intrusion into the streetscape. Only a small number of seats were fitted inside the vehicle, mainly to help visitors orientate themselves or to sit down when talking to GMPTE officials. A public event to promote the Metrolink project took place at the Birchfields Road site on 21 July 1988 and further open weekends were held in November of that year when preserved Manchester single deck tram 765 was also on display alongside the mock-up. Car 765 had been restored at Birchfields Road a number of years beforehand, and on this occasion it was en route from Blackpool to its home at the Heaton Park Tramway. Unfortunately attendance at the two open weekends was very poor,

Above: Docklands Light Railway P86 car 11 is seen near the temporary platform at Debdale Park on one of a series of demonstration runs on part of the Fallowfield Loop on 21 March 1987. *Keith Fender*

Above & below: Two views of the Metrolink tram mock-up at Birchfields Road bus garage, taken during a public event on 21 July 1988. *David Holt (2)*

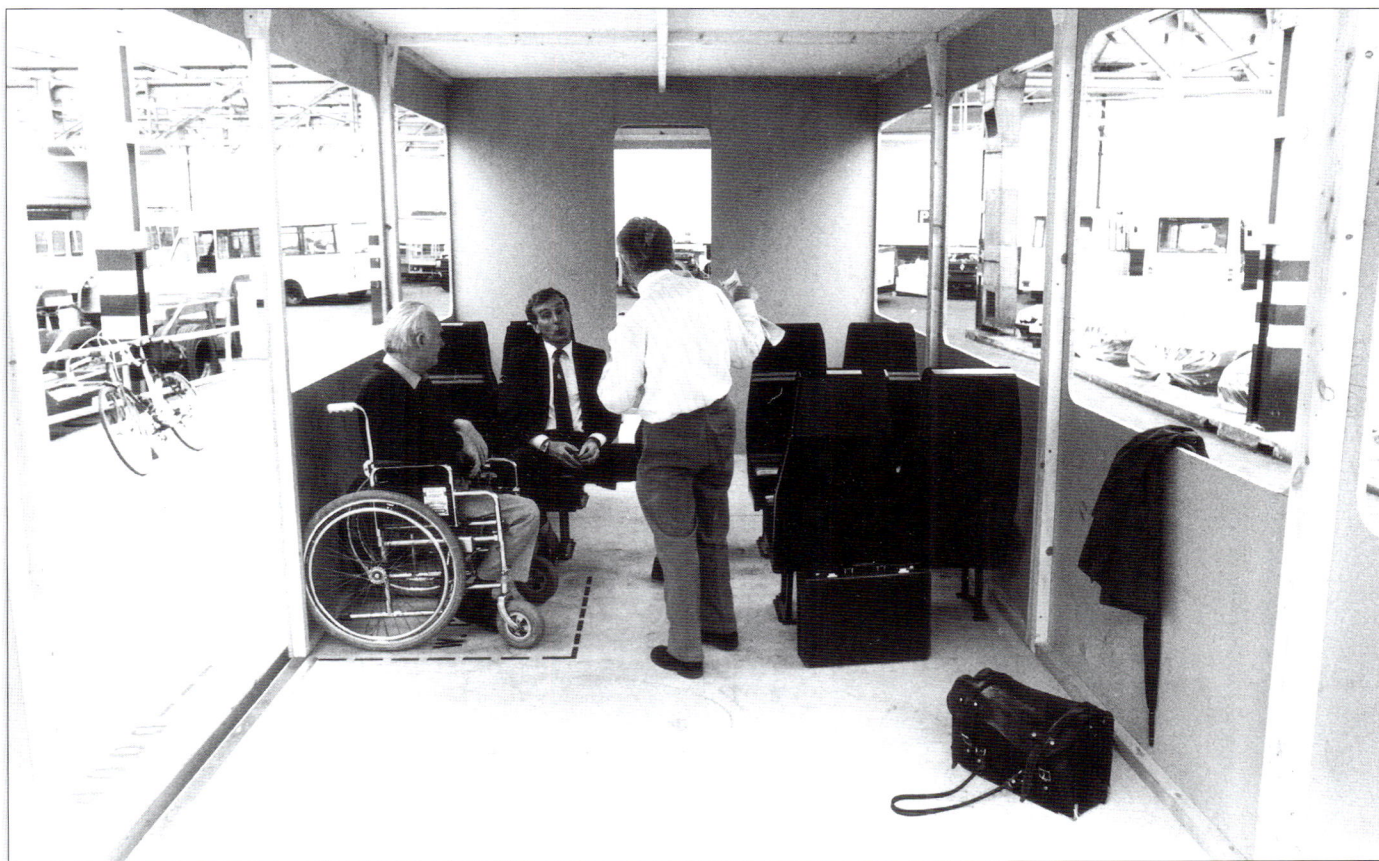

although it is reported that one visitor, a mother with a baby buggy, was asked what she thought of the gap between the platform and the tram, to which she answered that she did not actually notice it!

On 24 October 1989 Minister of State for Transport Michael Portillo announced that Section 56 funding would be available for Metrolink Phase 1, and in March and April 1990 another mock-up, this time built by Metrolink's chosen vehicle supplier Firema of Italy, was put on public display alongside a

Above: Mock-up T68 car 1000, named "The Larry Sullivan" after Councillor Larry Sullivan and now on display at the Greater Manchester Museum of Transport.

Below: The interior of mock-up car 1000. It is very similar to the production series T68s and T68As. *Robert Pritchard (2)*

Above & below: Two views of St Peter's Square during construction work in June 1991, showing Manchester Central Library and the Midland Hotel (above) and looking towards Piccadilly Gardens (below). On the right can be seen the cenotaph, which has since had to be resited to make way for the enlarged St Peter's Square tram stop. *Paul Abell (2)*

Left: Construction work in progress on London Road in June 1991, looking towards Aytoun Street and Piccadilly Gardens. The entrance to Piccadilly station undercroft is just off the picture to the right.

Below: Work in progress on Piccadilly Gardens tram stop in June 1991. On the left is Piccadilly bus station, with a Bee Line bus nearest the camera and two buses belonging to Manchester's then main bus operator GM Buses beyond it. *Paul Abell (2)*

specially built platform beneath a railway arch near Piccadilly station. This mock-up was built to a design closely resembling that of the T68 trams. It is now numbered 1000 and can be seen at the Greater Manchester Museum of Transport on Boyle Street, Cheetham Hill.

CONSTRUCTION STARTS

Street tramway construction was launched in a ground-breaking ceremony on 5 April 1990, and on 5 June Michael Portillo's successor Roger Freeman ceremonially installed the first grooved rails in

GET TO KNOW YOUR MET.

Britain's most advanced passenger transport system.   Get going. Get the MET

METROLINK

the city centre. To enable tracklaying to take place, and to avoid conflict with tram movements once the trams started running, all road traffic including buses had to be diverted except for buses on Mosley Street (inbound) and the traffic management system in the city centre completely redesigned. In the early days of bus deregulation this was no easy matter. GMPTE owned and managed the bus station facilities at Piccadilly Interchange and thus had a degree of control over the routing of bus services but still had to liaise with around 23 operators. One result of the rerouting of traffic to avoid conflict with the tramway was the removal of a road sign at the corner of St Mary's Gate and Deansgate giving directions both to the centre of Manchester and to Carlisle.

GMPTE had originally envisaged the new light rail system as part of an integrated transport network where the trams would complement the city's bus and heavy rail services, as such integration had long been regarded as an essential part of a truly efficient local transport system. The Tyne & Wear Metro in its first six years of operation was a good example of such a system, with feeder bus services and inter-available tickets covering both modes. However, just three months after GMPTE had submitted its Section 56 application for Government funding for Phase 1 of the light rail scheme in July 1985, October of that year saw the passing of the Transport Act 1985 under which all bus services in mainland Britain outside Greater London were to be deregulated with effect from 26 October 1986. This meant that applicants for Section 56 grants now had to show that their proposed scheme would be compatible with a deregulated environment. It also effectively ruled out any integration between buses and either light or heavy rail services, instead allowing bus operators to compete with other modes. However, although a degree of competition between buses and light rail has existed ever since Metrolink opened, experience has shown that passengers making longer distance journeys tend to prefer to travel by tram or train, meaning that in practice many of the city's bus services continue to cater mainly for flows not served by Metrolink or heavy rail.

During construction work in the city centre, old tram tracks from the city's original tramway were unearthed in a number of locations including Aytoun Street, Market Street and St

Peter's Square. Further tram tracks were discovered during the construction of the Second City Crossing. Much of the original tram track was simply tarmacked over rather than being lifted, and some transport historians have argued that the first generation tramways in Manchester and other UK cities could and should have been retained and modernised. One can only speculate about the more extensive network that might now be in operation if past transport policy had been more enlightened!

THE OLD ORDER ON THE BURY AND ALTRINCHAM LINES

Above: A Class 504 EMU, with car M77170 bringing up the rear, calls at Crumpsall with a Victoria–Bury service on 15 October 1988, just under three years before the end of heavy rail operation on the Bury line. The 504s never had unit numbers as such, only individual car numbers. *Gavin Morrison*

Above: A Class 504 EMU, with car M77167 leading, is seen near Cheetham Hill as it approaches Collyhurst Tunnel with a Victoria–Bury service, also on 15 October 1988. *Gavin Morrison*

On the Bury and Altrincham lines north of Victoria and south of Castlefield, the existing heavy rail tracks were largely retained but had to be brought up to the condition set out in the terms of the contract. Work carried out on the tracks here included rerailing, reballasting and resleepering. Some sections of track were already completely relaid by British Rail before the changeover. Both lines also had to be fully resignalled, and end-on links onto the street section were established with the Bury line at Victoria and with the Altrincham line at Windmill Street adjacent to the G-Mex (now Manchester Central) Convention Centre. Several bridges and viaducts had to be refurbished and a new depot was built at Queens Road on the Bury line. Track remodelling was carried out at Altrincham to separate the Metrolink and BR tracks. 18 of the existing BR stations were improved and adapted for disabled access by installing lifts or ramps, although in many cases the existing station buildings were retained but the stations were destaffed.

OWNERSHIP CHANGES

The Metrolink infrastructure and tram fleet has always been publicly owned by Transport for Greater Manchester/GMPTE. The PTE itself involved the public at every stage of the development of the Metrolink project, consulting extensively with user group representatives throughout the planning stages. Because of this, and by undertaking a series of study tours of light rail systems in mainland Europe and North America as mentioned above, also visiting the National Tramway Museum at Crich, the promoters of the scheme had thoroughly clued themselves up about tramways and were clearly in a strong position to see the project through to completion. However, rather than opt for a conventional client/contractor relationship, the Government of the day preferred to involve the private sector in the project and forced GMPTE to stand aside while contractors fresh from bypasses and heavy segregated railways were let loose in the city centre streets. It could be said that the Government saw Metrolink as a flagship project for railway privatisation.

For its first five years of operation, between 1992 and 1997, Metrolink was operated as a concession by Greater Manchester Metro Ltd, a private company formed by the GMA consortium comprising GEC Alsthom, John Mowlem and AMEC in 1989 to design, build and operate the Metrolink system. On 26 May 1997 Serco Metrolink, a wholly owned subsidiary of Serco, took over the maintenance and operation of the system for the next ten years. In 2007 Stagecoach was awarded the new Metrolink franchise, which started on 15 July that year. Unlike Serco, Stagecoach did not own the concession but ran it on a fixed-term management contract. As part of the deal GMPTE did not allow Stagecoach to apply its corporate livery to the tram fleet unlike on Sheffield Supertram.

However, although Stagecoach's Metrolink contract was to have run until July 2017, on 1 August 2011 RATP Dev UK Ltd, a subsidiary of Paris transport operator RATP, bought the concession from Stagecoach in a surprise move. Since July 2017 KeolisAmey, a partnership between Keolis and Amey, has operated Metrolink in a contract that runs until 2027. RATP Dev was one of the shortlisted bidders for the 2017–27 concession; however, interim Mayor of Greater Manchester Tony Lloyd said the decision to award the contract to KeolisAmey was based on the consortium's plans for the network including creating over 300 new jobs and increasing the number of staff on duty during evenings and weekends.

Above: Class 304 EMUs were the mainstay of Altrincham line services for 20 years from the conversion of the line's electrification system from 1500 V DC to 25 kV AC in 1971 until its closure for conversion to light rail in December 1991. Here unit 304 020 awaits departure from Altrincham with an Alderley Edge service on 18 May 1991. Another unidentified Class 304 unit can be seen on the left. *Gavin Morrison*

CHAPTER 5:

METROLINK ROUTE BY ROUTE

Visitors to Manchester will most likely have their first encounter with Metrolink at Victoria station, in Piccadilly station undercroft or in the city centre, although the network also serves Altrincham, Navigation Road, Manchester Airport and Rochdale stations (and Deansgate-Castlefield, which is located adjacent to Deansgate heavy rail station) where passengers may choose to change onto Metrolink or see the trams from trains running via these stations. All directions are given assuming that the reader is facing the direction of travel.

PHASE 1: ALTRINCHAM–BURY

Work on the conversion of the Bury line started in July 1991. The old heavy rail line closed in two stages, Victoria–Crumpsall on 13 July and Crumpsall–Bury on 16 August. Then on 24 December 1991 the Altrincham line closed for conversion to light rail. Phase 1 of Metrolink opened in four stages:

- 6 April 1992: Victoria–Bury
- 27 April 1992: Victoria–G-Mex (now Deansgate-Castlefield)
- 15 June 1992: G-Mex–Altrincham
- 20 July 1992: Piccadilly Gardens–Piccadilly

Metrolink was officially opened by Queen Elizabeth II on 17 July 1992.

Opened by the Lancashire & Yorkshire Railway in 1879, the Bury line had used a unique 1200 V DC side contact third rail electrification system since it was electrified by the L&Y in 1916, and had been worked by a purpose-built fleet of 2-car Class 504 EMUs since 1959. These operated only between Bury and Victoria, where

Below: Taken using a drone, M5000 cars 3031 and 3016 cross the River Irwell near Elton Reservoir shortly after departure from Bury with the 15.50 to Piccadilly on 12 September 2020. *Tom McAtee*

TABLE 1: DATES OF COMMENCEMENT OF PUBLIC SERVICE BY SECTION

Date	Section
6 April 1992	Bury–Victoria
27 April 1992	Victoria–G-Mex*
15 June 1992	G-Mex*–Altrincham
20 July 1992	Piccadilly–Market Street/Mosley Street
6 December 1999	Cornbrook–Broadway
21 July 2000	Broadway–Eccles
20 September 2010	MediaCityUK branch
7 July 2011	Trafford Bar–St Werburgh's Road
13 June 2012	Victoria–Oldham Mumps
16 December 2012	Oldham Mumps–Shaw & Crompton
11 February 2013	Piccadilly–Droylsden
28 February 2013	Shaw & Crompton–Rochdale railway station
23 May 2013	St Werburgh's Road–East Didsbury
9 October 2013	Droylsden–Ashton-under-Lyne
27 January 2014	Oldham town centre route
31 March 2014	Rochdale railway station–Rochdale Town Centre
3 November 2014	St Werburgh's Road–Manchester Airport
6 December 2015	Victoria–Exchange Square
26 February 2017	Exchange Square–St Peter's Square
22 March 2020	Pomona–intu Trafford Centre

*Now Deansgate-Castlefield.

Map labels

Bury, Radcliffe, Whitefield, Besses o' th' Barn, Prestwich, Heaton Park, Bowker Vale, Crumpsall, Abraham Moss, Queens Road, Victoria, Exchange Square, Shudehill, Market Street, Piccadilly Gardens, Piccadilly, New Islington, St Peter's Square, Deansgate-Castlefield, Cornbrook

Milnrow, Kingsway Business Park, Newbold, Rochdale Railway Station, Rochdale Town Centre, Shaw and Crompton, Newhey, Derker, Oldham Mumps, Oldham Central, Oldham King Street, Westwood, Freehold, South Chadderton, Hollinwood, Failsworth, Newton Heath and Moston, Central Park

Ashton-under-Lyne, Ashton West, Ashton Moss, Audenshaw, Droylsden, Cemetery Road, Edge Lane, Clayton Hall, Velopark, Etihad Campus, Holt Town

ZONE 1, ZONE 2, ZONE 3, ZONE 4

Broadway, Harbour City, Anchorage, Exchange Quay, Ladywell, Langworthy, Weaste, MediaCityUK, Eccles, Salford Quays, Pomona, Wharfside, Imperial War Museum, Old Trafford, Trafford Bar, Firswood, Chorlton, Village, Parkway, Stretford, St Werburgh's Road, Withington, Barlow Moor Road, West Didsbury, Wythenshawe Park, Sale Water Park, Northern Moor, Baguley, Moor Road, Roundthorn, Martinscroft, Benchill, Crossacres, Wythenshawe Town Centre, Robinswood Road, Peel Hall, Shadowmoss, Manchester Airport, intu Trafford Centre, Barton Dock Road, Dane Road, Sale, Burton Road, Didsbury Village, East Didsbury, Brooklands, Timperley, Navigation Road, Altrincham

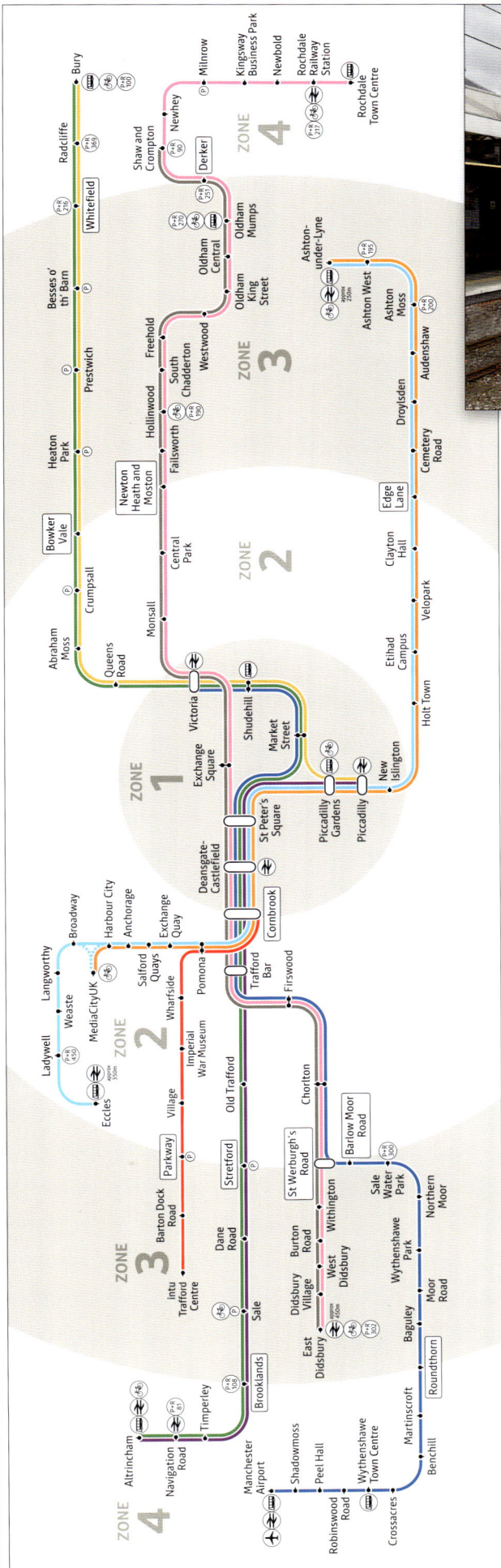

Above: T68 car 1021 awaits departure from Altrincham with a Piccadilly service on 3 May 2010. The heavy rail platforms used by Manchester–Chester trains can be seen on the right. *Robert Pritchard*

they used dedicated terminal platforms on the site of the present day Metrolink station, and the Bury line was thus a completely self-contained operation in British Rail days.

The Altrincham line was opened by the Manchester, South Junction & Altrincham Railway in 1849 and was previously part of the 25 kV AC electrified network, and had been worked mainly by Class 304 slam-door EMUs since its conversion in 1971 from its original 1500 V DC electrification system inaugurated in 1931. In 1500 V DC days electric operation on the Altrincham line was restricted to London Road (later renamed Piccadilly)–Altrincham and latterly Oxford Road–Altrincham, then the 25 kV enabled through trains to run via Oxford Road and Piccadilly to a number of destinations south of Manchester including Hazel Grove (after its electrification in 1981), Alderley Edge via Styal, Crewe via Wilmslow, and Stafford via Stoke-on-Trent. Manchester–Chester via Northwich DMUs running via the Mid-Cheshire Line also ran via Sale (not calling at most of the intermediate stations) until May 1989 when they were diverted via Stockport. Manchester Oxford Road–Liverpool via Lymm and Warrington Bank Quay (Low Level) trains also shared the line as far as Timperley until September 1962.

Altrincham station forms part of a bus/rail interchange established in 1976 as part of the preparations for the aborted Picc-Vic Tunnel scheme. Today, Platforms 1 and 2 are used by Metrolink and Platforms 3 and 4 by Northern's Manchester–Northwich–Chester trains. Shortly after departure from Altrincham on the right is a crossover, which is mainly used to deliver stone ballast and other materials to the Metrolink lines. Just before the first station, Navigation Road, both the Metrolink and the Network Rail tracks become single track. This means that the former Altrincham-bound platform at this station is used by Mid-Cheshire Line DMUs (and freight trains also pass it in both directions) and the former Manchester-bound platform is used by Metrolink trams travelling in both directions. Just after Navigation Road the line

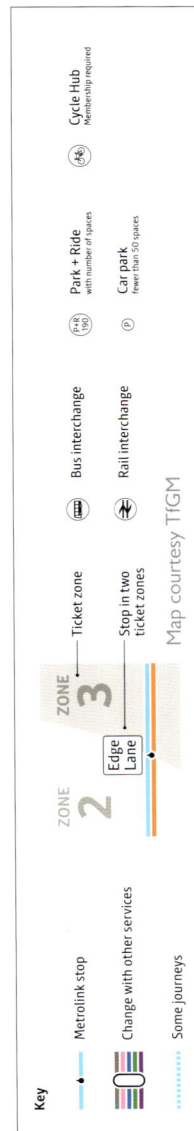

Key

- Metrolink stop
- Change with other services
- Some journeys

Ticket zone
Stop in two ticket zones

Cycle Hub — Membership required
Park + Ride — with number of spaces
Car park — fewer than 50 spaces
Bus interchange
Rail interchange

Map courtesy TfGM

Left: Most of the stations on the Altrincham and Bury lines have lost many of their heavy rail era buildings and fixtures, but Sale retains its elegant looking canopies on both platforms. M5000 car 3046 arrives with an Altrincham–Piccadilly service on 14 May 2014. *Steve Hyde*

Right: T68 car 1016 calls at Stretford with an Altrincham–Bury service on 17 April 2012. In the background can be seen the heavy rail era street level station building, still in situ and owned by TfGM but rented out to other users since the conversion of the line to Metrolink. The heavy rail era overhead wire gantries are also still in evidence in this view. *Alan Yearsley*

Left: M5000 car 3062 arrives at Deansgate-Castlefield with a Piccadilly-bound service on 18 September 2014. *Robert Pritchard*

Above: M5000 car 3050 passes Manchester Central (formerly G-Mex) Convention Centre as it descends from the viaduct across Great Bridgewater Street and heads towards St Peter's Square with a Piccadilly-bound service on 18 September 2014. *Robert Pritchard*

to Stockport via Northenden Junction takes a sharp curve to the right at Deansgate Junction, after which we pass under the trackbed of the former Northenden–Glazebrook line. The Bridgewater Canal then runs alongside us on our left as we pass through the next three stops: Timperley, Brooklands and Sale, after which the canal swings slightly to the left but still runs close to the line.

Shortly after the next stop, Dane Road, the line passes under the Manchester Outer Ring Road. This is immediately followed by Sale Water Park on the right, and we then cross over the River Mersey

as it crosses under the canal, which remains in close proximity to the line until shortly after Stretford. On the left at Hawthorn Lane is an aqueduct built to the same design as James Brindley's original Barton aqueduct. The line runs beneath the A5145 Edge Lane just before Stretford stop, then curves off to the right. The next stop, Old Trafford, is located adjacent to the famous Old Trafford Cricket Ground, home of Lancashire County Cricket Club, which can be seen on the left. Old Trafford Football Ground, home of Manchester United FC, is located nearby.

We then pass Trafford depot on the right, after which the line from East Didsbury and Manchester Airport trails in just before Trafford Bar stop. Opened in 2011, Trafford is the second depot to open on Metrolink (the first being at Queens Road on the Bury line). Rather confusingly, in heavy rail days the present day Old Trafford and Trafford Bar stops were named Warwick Road and Old Trafford respectively. This

Left: M5000 car 3001 calls at St Peter's Square with a Piccadilly-bound service on the evening of 9 January 2010 as T68A car 2003 heads away from the camera on the left. On the right can be seen Manchester Central Library. *Robert Pritchard*

Left: M5000 car 3009 passes the now-closed southbound-only Mosley Street stop with a Piccadilly-bound service on 16 February 2012. *Robert Pritchard*

Below: T68 cars 1017 (nearest the camera) and 1023 head along Aytoun Street with a Piccadilly-bound service in the spring of 1995. *Peter Fox*

often misled cricket match goers into alighting at Trafford Bar (then Old Trafford) by mistake, not realising that Warwick Road was actually adjacent to the ground!

After Trafford Bar, the line runs through the short Old Trafford Tunnel, then passes beneath the Manchester–Warrington–Liverpool main line via an underpass that was built at the site of the former Old Trafford Junction where the Altrincham line used to diverge from the line into Manchester Central. The line from Eccles and MediaCityUK then trails in from the left and right just before we reach Cornbrook, the only newly built stop between Altrincham and central Manchester that did not open at the start of Metrolink operations. Cornbrook opened on 6 December 1999, initially only for transfer between Bury/

Altrincham and Eccles line trams. Street level entry and exit has been allowed since 3 September 2005.

The line then follows the alignment of the former Cheshire Lines Committee line into Manchester Central as far as Deansgate-Castlefield, passing over Castlefield Canal Basin just before reaching Deansgate-Castlefield stop, which is located adjacent to Manchester Central Convention Centre (originally opened as G-Mex in 1986 and used during the Covid-19 pandemic as a Nightingale Hospital) in the former Central station. The Manchester–Warrington–Liverpool main line and Deansgate heavy rail station (known as Knott Mill & Deansgate until 1971) are on our right, with Deansgate station being linked from the tram stop by a footbridge over Whitworth Street West.

Left: M5000 car 3001 arrives at Piccadilly station undercroft with an Altrincham service on 9 January 2010.

Below: M5000 car 3010 heads along High Street between Market Street and Shudehill stops with a Bury service on 17 April 2012. The destination appears to read Bury via Woodlands Road, as trams called at the now closed Woodlands Road stop only between around 10.00 and 16.00 on weekdays from the opening of the nearby Abraham Moss stop on 18 April 2011 until the closure of Woodlands Road on 16 December 2013.

The station still boasts its original Manchester, South Junction & Altrincham Railway street level building, which dates from 1896 and bears the name Knot Mill Station above the entrance.

After Deansgate-Castlefield the line curves off to the left onto a long ramping viaduct with the Convention Centre on the left and the Bridgewater Hall concert venue on the right. Trams descend from the viaduct onto Lower Mosley Street, then continue to St Peter's Square where one of the city's

Above: The rebuilt Metrolink station at Victoria has three tracks and four platform faces, the centre road normally being used by trams starting and terminating there. From left to right, M5000 car 3054 is seen with a Rochdale via Oldham service, car 3017 forms a Manchester Airport service (starting from the bay platform here) and cars 3003 and 3039 form a Bury–Piccadilly service on 19 April 2019. *Robert Pritchard (3)*

Above: T68 car 1021 leaves Market Street with a Bury–Piccadilly service on 29 April 2007.

Left: M5000 car 3020 arrives at Shudehill with a Bury–Piccadilly service (again via Woodlands Road) on 17 April 2012. *Robert Pritchard (2)*

number of services increased and a platform upgrade at this stop to accommodate pairs of M5000 cars was not deemed economic so it was closed on 18 May 2013.

Under the normal timetable, trams from Altrincham run alternately to Piccadilly and direct to Bury (and likewise trams from Bury run alternately to Piccadilly and to Altrincham). Under the reduced timetable in operation during 2020–21 due to the Covid-19 pandemic direct Altrincham–Bury trams did not operate, however. Altrincham–Piccadilly trams turn right into the island

best known landmarks, Manchester Central Library, is on the left. St Peter's Square stop consists of two island platforms, as it is served by trams running both on the original route through the city centre (used by Altrincham–Bury trams) and by those that use the Second City Crossing to Victoria, which curves off to the left past Manchester Town Hall. Altrincham/Bury line trams continue straight on along Mosley Street to Piccadilly Gardens. Until 2013 there was a stop at Mosley Street, just short of Piccadilly Gardens, for southbound trams only, but this was controversially found to cause congestion as the

platform stop at Piccadilly Gardens, then continue along Aytoun Street to Piccadilly station tram stop located in an undercroft beneath the main station, which can be reached from the concourse by escalator, lift or fixed staircase or directly from the adjacent streets. Trams starting from Piccadilly take this route in the reverse direction as far as Piccadilly Gardens, then turn left if bound for Altrincham or right then immediately left along Market Street if heading for Bury. Direct Altrincham–Bury trams bypass Piccadilly Gardens on the right and then turn left onto Market Street, soon arriving at Market Street

Left: T68 car 1003 departs from Crumpsall with a Bury–Altrincham service on 8 October 2008. Crumpsall stop was rebuilt between 2017 and 2019, including provision of a new turnback siding and third platform that will eventually be used by terminating trams from the Trafford Park Line. *Robert Pritchard*

Below: An engineering possession on the Bury line during the summer of 2007 saw the replacement of the heavy rail era track and sleepers. Three EWS "cut down cab" Class 08 shunters (08993–995) had to be used on the works trains owing to restricted clearances, such as this one at Bowker Vale with 08993 on 9 July 2007. At this time the heavy rail era station buildings and canopies were still in situ. *John Myddelton*

stop, also consisting of a single island platform opened in August 1998. Until this time, Market Street was a single platform served only by Bury-bound trams, with southbound trams calling at the now demolished High Street stop.

The line then curves to the right onto High Street (where the original High Street stop was located), passing the heart of Manchester's main city centre retail quarter with the Arndale Centre on the left extending as far as the next stop, Shudehill Interchange. Completed in 2002, Shudehill tram stop opened on 31 March 2003 followed by the bus part of the interchange on 29 January 2006. Construction of the bus station started in the late 1990s but was halted until 2003 because of a series of legal disputes over the ownership of the land.

After Shudehill, the line runs steeply downhill along Balloon Street, with Co-

Left: M5000 car 3026 calls at Bowker Vale with a Bury service on 17 October 2018 (right) as a pair of M5000s formed of cars 3014 and 3027 bound for Piccadilly departs (left) with 3027 bringing up the rear. The heavy rail era station buildings had been demolished and replaced by new shelters by the time this photo was taken. The new design of tram stop provides less protection from the elements but is also less prone to vandalism than the old structures. *Steve Hyde*

operative buildings on both sides, to Victoria station where the Second City Crossing trails in on our left. Victoria station underwent a £44 million renovation between April 2013 and August 2015, including restoration of the station's period features, a new roof, and an enlarged Metrolink station.

Above: T68 car 1022 calls at Whitefield with a Bury–Altrincham service on 8 October 2008. Seen on the right is one of the original Thorn EMI ticket vending machines. *Robert Pritchard*

Right: T68A car 2006 calls at Exchange Quay with an Eccles–Piccadilly service on 12 July 2012. *Andrew Thompson*

Apart from Piccadilly station undercroft, Shudehill–Victoria is the only section of the city centre Metrolink network that is not part of an original Manchester Corporation tramway alignment.

When Metrolink first opened, Victoria station had just one wide island platform used by trams running to and from Bury. Between 2013 and 2015 the station was extensively rebuilt to accommodate the expanded tram service and a new overall roof was constructed over the Metrolink part of the station to replace the previous structure. The present day Victoria Metrolink station consists of three tracks with two island platforms between them. There are four platform faces lettered A, B, C and D. Platforms A (southbound) and D (northbound) are used by through trams and the single track between Platforms C and D is mainly used by terminating trams. Access to both Metrolink platforms is via a short ramp from the main concourse.

On leaving Victoria, the line climbs up Miles Platting bank and runs alongside the heavy rail tracks for a short distance, then dives under the main line by way of Collyhurst Tunnel. This is soon followed by Irk Valley Junction, where the Rochdale line diverges to the right and the Bury line then passes Queens Road depot on the left, Metrolink's original depot. Until 2013 the depot was served only by a staff halt, but this has now been replaced by a proper public tram stop. The line then enters a deep cutting that takes it to the now abandoned Woodlands Road stop, closed in 2013, after which we climb towards its replacement stop, Abraham Moss, which opened in 2011 to serve new housing and business developments in the area and features a park & ride facility. From here it is only a short distance through

Above: Refurbishment of the standard T68 cars, including the fitting of retractable couplers and covered bogies, enabled them to run on the Eccles line, which had been worked exclusively by the T68As for its first few years of operation. Here T68 car 1005 passes the junction for the MediaCityUK branch, then under construction, with an Eccles service on 3 May 2010. *Robert Pritchard*

another cutting to the next stop, Crumpsall, after which the line ascends onto an embankment and heads in a north-westerly direction as it crosses over the A576 Middleton Road and reaches Bowker Vale. Visitors to Manchester's heritage tramway in Heaton Park can alight here or remain on board to the park's namesake stop, which is reached immediately after Heaton Park Tunnel.

The line then climbs out of a cutting and onto an embankment once again, which leads us to the next stop at Prestwich, followed by the famous Besses o' th' Barn bridge, which consists of a reinforced concrete beam structure shaped like an inverted "T" with the northbound track on one side and the southbound track on the other. This bridge was installed when the M60

Left: M5000 cars 3118 (nearest the camera) and 3090 pause at the MediaCityUK terminus with an Eccles–Ashton service on 18 July 2020. *Alan Yearsley*

Manchester Outer Ring Road was built in the late 1960s, and enables the A665 Bury Old Road to pass underneath the Metrolink line and over the M60. After calling at the island platform station at Besses o' th' Barn we descend from the embankment to ground level by the time we reach Whitefield. There is then a short tunnel beneath the A56 Bury New Road, after which the line emerges into a long deep cutting that was a source of no end of engineering problems during the construction of the line by the Lancashire & Yorkshire Railway in the 1870s. The line then passes under the A665 Radcliffe New Road and climbs up onto a viaduct over the River Irwell just before reaching the last intermediate stop at Radcliffe. Shortly afterwards the Manchester, Bolton & Bury Canal runs close to the line on our left until we reach the only level crossing on the Bury line and then cross over the River Irwell for a second time. This level crossing, located on the A56 road, is at the site of a new stop at Buckley Wells proposed by GMPTE in 2003 but not progressed any further to date.

As we approach Bury the Western Pennine Moors can be seen in the distance, a far cry from the relatively flat Cheshire Plain at the Altrincham end of the line! On the left the little-used spur linking Metrolink with the original alignment of the Bury line (now the East Lancashire Railway) diverges, and the line then passes under the Bury–Heywood section of the ELR, which crosses over the Metrolink line on a bridge known as the "Ski Jump", built close to the site of the old Bury Knowsley Street station on the former Bolton–Heywood–Rochdale line. We then arrive at Bury Interchange, the northern terminus of the line since 1980 when it replaced the nearby Bury Bolton Street station (now used by the ELR).

Right: M5000 car 3013 traverses the crossover at Broadway with an Eccles–Piccadilly via MediaCityUK service on 26 February 2012. At this time all Eccles line services ran via MediaCityUK. *Steve Hyde*

PHASE 2: ECCLES

During the 1990s the site of the former Manchester Docks, now known as Salford Quays, was redeveloped as a major residential and business quarter. As the area had poor public transport provision, a Parliamentary Bill for an extension to Salford Quays was deposited in November 1987 and received Royal Assent in April 1990. This line had originally been proposed in 1985 by David Holt, the author of Platform 5 Publishing's original 1992 Metrolink book, after the area had been completely cleared of Docks buildings.

Work on the Salford Quays and Eccles extension began in April 1997 following the signing of the contract between GMPTE and Altram, a consortium comprising Ansaldo, Serco and John Laing. The line was officially opened as far as Broadway by Prime Minister Tony Blair on 6 December 1999. Services to Eccles started on 21 July 2000, with the whole line being officially declared open in a ceremony by Anne, Princess Royal on 9 January 2001.

After diverging from the Altrincham line at Cornbrook, the line crosses over the Bridgewater Canal and the Manchester Ship Canal

Above: T68A car 2005 crosses over from the on-street section to the reserved track at the junction of Foster Street as it approaches Weaste with an Eccles service on 8 October 2008. *Robert Pritchard*

Above: Shortly after leaving Eccles, T68A car 2005 approaches Ladywell with a Piccadilly service on 3 May 2010 having just passed beneath the Ladywell roundabout. *Robert Pritchard*

by way of the 650 m long Pomona Viaduct. The first stop, Pomona, an island platform, is located on this viaduct south of the Ship Canal. Immediately after Pomona the new Trafford Park line continues straight on, and the Eccles line turns sharp right and takes a curvaceous route through the Salford Quays development, which had already been built before the tramway so the line had to be constructed around the existing buildings instead of the other way round. This means that journeys on the Eccles line are somewhat slower than if it could have been planned at the same time as the residential and office buildings. A notable feature of the Eccles line is the use of lawn track with grass growing between the rails (and between the two running lines) on the section between Anchorage and Broadway, a feature also found on many tramways in mainland Europe particularly on the new tramways in France such as Bordeaux.

There are five more stops on the reserved track section: Exchange Quay, Salford Quays, Anchorage, Harbour City, and Broadway. Just after Harbour City the short single track branch to MediaCityUK, opened on 20 September 2010, continues straight ahead while the Eccles line curves off to the right with the other half of the MediaCityUK triangle junction then trailing in on the left. After Broadway, the line runs on-street along South Langworthy Road and then the A57 Eccles New Road following the former Salford Corporation Tramways alignment all the way to Eccles with three more intermediate stops at Langworthy, Weaste and Ladywell. Between Ladywell

and Eccles the line dives under the Ladywell Roundabout, then runs alongside Regent Street to the terminus at Eccles. Just beyond the tram stop is Eccles bus station, nowadays known as Eccles Interchange, and by turning right and taking a short walk through the pedestrianised shopping area Eccles rail station can be reached.

PHASE 3

In the early 2000s GMPTE and the Association of Greater Manchester Authorities (AGMA) promoted the Phase 3 extension project, nicknamed the "Metrolink Big Bang" and consisting of the takeover of the existing Oldham Loop heavy rail line to Rochdale via Oldham and three new lines to Ashton-under-Lyne via Droylsden and to East Didsbury and Manchester Airport via St Werburgh's Road. These

Right: 66093 leads Pathfinder Railtours' "The Hacienda" railtour from Swindon into the Dean Lane Waste Terminal (66503 was on the rear) as M5000 car 3031 heads away from the camera bound for Shaw & Crompton on 7 March 2020. *Andy Chard*

Above: M5000 car 3036 departs from Central Park with a St Werburgh's Road service on 18 August 2012. Note the impressive roof of Central Park tram stop.

Left: 3021 departs from Central Park with an Oldham Mumps service on 18 August 2012. *Robert Pritchard (2)*

rising costs. This led GMPTE to launch a "Get Metrolink Back on Track" campaign, with T68 tram 1015 being outshopped in a special purple livery for publicity purposes. Eventually, in July 2006 the Government gave conditional funding approval for Phase 3a, with the funding package being approved in May 2008, then in May 2009 AGMA agreed to create a £1.5 billion Greater Manchester Transport Fund including Metrolink Phase 3b and the Second City Crossing.

OLDHAM AND ROCHDALE

The Rochdale via Oldham line was opened by the Lancashire & Yorkshire Railway in three stages: from Middleton Junction to Oldham Werneth in 1842, then on to Oldham Mumps in 1847 and finally to Rochdale East Junction in 1863. In 1880 the more direct route from Thorpes Bridge Junction via Hollinwood opened, and the original Middleton Jn–Oldham Werneth line closed in 1963. The line is commonly referred to as the Oldham Loop; however, strictly speaking the original Middleton Jn–Oldham Werneth line was part of the Oldham Loop but the route via Hollinwood as far as

were divided into two distinct phases, with Phase 3a comprising the Oldham Loop conversion, the South Manchester Line as far as St Werburgh's Road and the East Manchester Line as far as Droylsden, and Phase 3b the Ashton, East Didsbury, Oldham and Rochdale town centre and Manchester Airport extensions.

In December 2002 the Government announced £520 million of funding for the £820 million Metrolink Phase 3 project, with the remaining £300 million coming from the private sector. However, in July 2004, the Government shelved funding for Phase 3 because of

Right: An exterior view of Central Park tram stop as seen from the Rochdale-bound platform.

Werneth is not, so Werneth is where the Oldham Loop proper begins.

Trains ran from Victoria as far as Thorpes Bridge Junction, where the Oldham line diverged to the right. In the last few years of heavy rail operation the line had four trains per hour, two calling at all stations and terminating at Shaw & Crompton, the other two calling only at Oldham Mumps, Shaw & Crompton and then all stations to Rochdale. Heavy rail services via Oldham ran for the last time on

Below: M5000 car 3042 departs from Freehold as it passes the now demolished Hartford mill with a St Werburgh's Road service on 19 April 2013.
Alan Yearsley (2)

Above: Work in progress on the demolition of the former railway viaduct in Oldham on 18 August 2012.
Robert Pritchard

3 October 2009, and the line then closed for conversion to Metrolink. On this day Class 156 DMU 156 466 was named "Gracie Fields" after the Rochdale-born singer and actress, almost 30 years to the day since her death in 1979, and Spitfire Railtours' "The Witch Way" railtour made two circuits of the Oldham Loop in the evening: one in a clockwise direction via Rochdale and Oldham steam-hauled by ex-LMS Stanier Class 5 4-6-0 45231 "The Sherwood Forester" and then in the other direction diesel-hauled by West Coast Railway Company's 37706.

After the closure, replacement buses were initially provided but these were soon withdrawn for lack of patronage. Instead most passengers chose to use nearby stations on adjacent lines such as Rochdale, Castleton, Mills Hill and Moston on the Calder Valley line and Ashton-under-Lyne on the Manchester–Huddersfield line. This led to severe overcrowding on both these lines, with reports of passengers being left behind on occasions. Because of this, GMPTE had to pay to reinstate the five Class 142 DMUs that the Department for Transport had stood down following the Oldham Loop closure.

Whereas the Altrincham and Bury lines only served the existing stations when first converted to light rail with no new stops being added, this was not the case on the Oldham Loop. Dean Lane station was renamed Newton Heath & Moston, which uses only the former Manchester-bound platform as the line was singled through the station to enable Network Rail to use the other line to access the Dean Lane Greater Manchester Waste Disposal facility. New stops were added at Monsall and Central Park on the new alignment between

Below: M5000 car 3002 awaits departure from the temporary terminus on the former railway alignment at Oldham Mumps with a St Werburgh's Road service on 18 August 2012. *Robert Pritchard*

Left: M5000 car 3054 with a Rochdale service on the original route through Oldham on the last day of operation via this alignment, 17 January 2014. This photo was taken near the site of the former stations at Oldham Central on the Oldham Loop and Oldham Clegg Street on the Oldham, Ashton & Guide Bridge Railway. *Steve Hyde*

to Rochdale station on 28 February 2013. Then on 27 January 2014 the new on-street route through Oldham town centre opened, ten days after the last trams ran via the old alignment, with a new Mumps tram stop to replace the existing one and two completely new stops at Oldham King Street and Central. Finally, on 31 March 2014 the Rochdale Town Centre extension opened.

The Rochdale line diverges from the Bury line at Irk Valley Junction, then passes through the short Victoria and Newton Heath, at South Chadderton, Freehold and Westwood between Hollinwood and Oldham (Freehold and Westwood stops effectively replacing Oldham Werneth station), and at Kingsway Business Park and Newbold between Oldham and Rochdale.

The first phase of the new Metrolink line opened as far as Oldham Mumps on 13 June 2012, with trams initially following the former railway alignment and using a temporary tram stop on the site of the former Mumps station (albeit at ground level, the original elevated station and overbridge having been removed in August 2010). Services were then extended to Shaw & Crompton on 16 December 2012 and

Right: M5000 cars 3007 and 3035 call at Oldham King Street on the new alignment through Oldham town centre with a Rochdale service on 30 October 2014. *Robert Pritchard*

Right: M5000 car 3090 stands in the bay platform for terminating trams at Shaw & Crompton with an East Didsbury service on 19 April 2019. This bay is on the site of the former signal box, and the present day tram stop platforms are located on the opposite side of the level crossing from the former heavy rail station. *Robert Pritchard*

Below: M5000 cars 3074 and 3079 on the scenic stretch between Shaw & Crompton and Derker with a Rochdale–East Didsbury service on 19 April 2019. *Robert Pritchard*

it runs through Westwood and then passes underneath the A627 Oldham Road before reaching Oldham King Street. From here the line runs on-street through Oldham town centre, with one more new stop at Oldham Central.

Just beyond the present day Oldham Mumps tram stop, the last stop in Oldham town centre, part of the trackbed of the temporary alignment used between 2012 and 2014 can be seen on the right. We then ascend onto the former railway alignment once again, running at broadly ground level as far as the next stop, Derker, which has a large park & ride facility on our right. Derker station originally opened

Queens Road Tunnel and emerges into a shallow cutting leading to the first stop, Monsall, after which the line runs on viaducts as far as Central Park stop, a futuristic looking building consisting of a curved copper and glass canopy suspended by a cable-tensioned steel structure. This was already built in 2005 in anticipation of the "Metrolink Big Bang" but delays to the project as described above meant that the facility lay unused until 2012.

After Central Park, the line crosses over the Calder Valley Line, then passes Newton Heath Depot located alongside Newton Heath & Moston tram stop. The line is now once again in a shallow cutting but ascends onto a viaduct just before Failsworth, shortly after which it crosses over the Rochdale Canal. This is followed by one more original station at Hollinwood, after which the line crosses over the M60 motorway and passes through three new stops at South Chadderton, Freehold and Westwood, the track remaining elevated almost all the way to Oldham town centre. Between Freehold and Westwood the line enters a shallow cutting and curves firstly to the right and then to the left, and as it does so the fenced-off original railway alignment can be seen on the right. The line then takes a curvaceous route as

Right: The now demolished Shaw & Crompton signal box, which was removed when the line closed for conversion to Metrolink. The turnback siding for terminating trams now occupies its former site. *Robert Pritchard*

Left: M5000 car 3004 approaches Newhey with a Rochdale service on 19 April 2019. *Robert Pritchard*

in 1985 to replace Royton Junction station (known as plain Royton from 1978), which closed in 1987 and was located just north of Derker. The Royton branch diverged on the left just after Royton Junction but this closed in 1966 and there is now very little trace left of the trackbed.

After the site of the former Royton Junction, the line has a more rural feel about it for most of the rest of the journey to Rochdale, albeit briefly passing through another built-up area just after the next stop, Shaw & Crompton. This is followed by two stops on the sites of already existing heavy rail stations at Newhey and Milnrow, and then after the first of the two new stops at Kingsway

Right: M5000 car 3002 awaits departure from Rochdale station (then the temporary terminus of the Rochdale line) with an East Didsbury service on 25 May 2013, against the backdrop of the iconic St John the Baptist Roman Catholic church, which was built in the Byzantine Revival style and modelled on the famous Hagia Sophia in Istanbul. *Robert Pritchard*

Below: M5000 car 3028 crosses the Rochdale Canal with a Rochdale service as it nears its destination on 25 May 2013. *Robert Pritchard*

Right: Car 3002 is seen again shortly after departure from Rochdale station on the same day as it curves onto the new dedicated right of way linking the on-street section with the former heavy rail alignment.

Below: Work in progress on the Rochdale town centre extension on 25 May 2013, with the sign on the left showing the route from the station to the new terminus.
Robert Pritchard

Business Park the line crosses over the Rochdale Canal and enters the outskirts of Rochdale as it passes the other new stop at Newbold. We then cross over and run alongside the Calder Valley line before arriving at Rochdale station. The trams use an island platform at right angles to the station entrance alongside St John the Baptist Roman Catholic church, built in 1927 in the Byzantine style modelled on the famous Hagia Sophia in Istanbul, Turkey. For the last leg of the journey through Rochdale town centre, the trams run on reserved track along Maclure Road, then take a sweeping curve along Drake Street where the dedicated right of way soon gives way to on-street running and becomes single track just before taking a sharp right turn into Rochdale Town Centre terminus, an island platform located alongside Rochdale Interchange, the town's main bus station. An intermediate stop on Drake Street was originally envisaged, but was dropped in 2011 during construction of the line because it was to be located only 200 m from the terminus and GMPTE officials believed that losing the additional stop would improve journey times. However, this decision was unpopular with local traders who had suffered disruption while the line was being built but did not believe that they would benefit from the new line without a stop that would be convenient for their customers.

Above: M5000 car 3081 heads along Drake Street in Rochdale town centre with a Rochdale–East Didsbury service on 19 April 2019.

Above: In 2019 M5000 car 3093 carried Vodafone advertising decals. It is seen arriving at Rochdale Interchange at the rear of an East Didsbury service on 19 April 2019, with car 3099 leading. *Robert Pritchard (2)*

THE OLD ORDER ON THE OLDHAM AND ROCHDALE LINE

Above: The old order on the Oldham Loop: 156 424 calls at Shaw & Crompton with a Manchester Victoria–Rochdale service on 8 October 2008, almost exactly a year before the line's closure for conversion to Metrolink. The present day tram stop is located on the opposite side of the level crossing to the former heavy rail station.

Above: A notice at Shaw & Crompton announcing the closure of the Oldham Loop for conversion to Metrolink, posted a year before this would take effect.

Above: 150 146 forming a Rochdale–Manchester Victoria service passes 150 277 on a Victoria–Rochdale service between Derker and Shaw & Crompton on 8 October 2008. *Robert Pritchard (3)*

EAST DIDSBURY

The South Manchester Line diverges from the Altrincham line at Trafford Bar and follows the trackbed of the former Manchester South District Railway route from Manchester Central to Stockport Tiviot Dale, opened by the Midland Railway in 1880. As well as local services, this line was also used by expresses to London St Pancras via Matlock. Local trains on the line were withdrawn and Tiviot Dale station closed in 1967, with the London expresses continuing until closure of Manchester Central station in 1969. The line between Chorlton and Cheadle Heath was then closed but freight trains continued to run on the Fallowfield Loop between Trafford Park Junction and Gorton Junction via Chorlton until 1988 when the track was lifted.

Part of the Fallowfield Loop line was used for the demonstration runs of a Docklands Light Railway car in 1987 (see Chapter 4). Once funding

Above: M5000 car 3026 after arrival at St Werburgh's Road with a terminating service on 18 August 2012. At this time St Werburgh's Road was being used as a temporary terminus for the South Manchester Line before services were extended to East Didsbury.

Left: M5000 car 3030 departs from St Werburgh's Road with an East Didsbury service on 29 February 2020. Just behind it the Manchester Airport line can be seen heading off to the left.

had been secured for the East Didsbury line, clearance work started in October 2008. The first stage from Trafford Bar to St Werburgh's Road was opened on 7 July 2011 followed by St Werburgh's Road–East Didsbury on 23 May

Right: M5000 car 3032 arrives at Didsbury Village with an East Didsbury–Rochdale service on 25 May 2013, just two days after the start of services between St Werburgh's Road and East Didsbury.
Robert Pritchard (3)

Above: M5000 car 3055 awaits departure from East Didsbury for Rochdale as the first ever service tram to leave the new terminus on 23 May 2013. Many of the group of intending passengers were to be seen on the first tram on each of the Phase 3 extensions as they opened. *Steve Hyde*

2013. GMPTE worked closely with Natural England to protect wildlife along the route, including a pledge to plant at least five young trees for every tree that had to be removed. In 2004 GMPTE applied for powers to build a further extension from East Didsbury to Stockport, but this was dropped when funding for Metrolink Phase 3 was shelved.

Upon leaving the Altrincham line we pass Trafford depot on the right, then enter a shallow cutting that takes us to the first stop at Firswood. The line then passes under the B5217 Manchester Road and curves to the left in a south-easterly direction as it passes the next stop, Chorlton, located on the site of the original Chorlton-cum-Hardy station on Wilbraham Road adjacent to a Morrisons supermarket.

Immediately after Chorlton stop, the line passes under Wilbraham Road and curves round towards St Werburgh's Road stop, where the former Fallowfield Loop line, now converted to a cycle path, curves off to the left. The Manchester Airport line then diverges to the right, after which we cross over the Chorlton Brook and then immediately pass under Mauldeth Road West. The line soon runs alongside Hough End Playing Fields on the left, which continues as far as the next stop, Withington. We then pass Albermarle Allotments on the left and Cavendish Road Park on the right just before Burton Road stop, after which the line reaches the large and popular Didsbury residential area. The first of three of its namesake stops is West Didsbury, which is reached immediately after passing under the junction of Palatine Road and Lapwing Lane. A shallow cutting then leads to the next stop, Didsbury Village, located on the south side of School Lane.

Shortly after Didsbury Village stop, part of National Cycle Network route 62, which runs from Fleetwood near Blackpool to Selby, North Yorkshire, crosses over and then runs alongside the tram route as far as the terminus at East Didsbury, which we reach just after passing under the Manchester Airport heavy rail line and the A34 (Kingsway). The cycle path ran along the trackbed and had to be diverted when the Metrolink line was built. It was initially only possible to plan for a single track line alongside the cycle path between School Lane and

East Didsbury, but in the event additional land became available and in January 2004 GMPTE was granted permission for a double track line.

The terminus at East Didsbury boasts a 302-space park & ride facility. It is located just the other side of Kingsway from a Tesco supermarket and a short walk from East Didsbury heavy rail station on the Manchester Airport line. Several other shops, bars, restaurants and entertainment venues are also located nearby.

ASHTON-UNDER-LYNE

Construction of the East Manchester Line to Droylsden and Ashton-under-Lyne started in November 2009, with the first test runs taking place between Piccadilly and Velopark in July 2011. Three days of trial passenger operation took place between Piccadilly and Droylsden on 8–10 February 2013, during which time local residents used free invitations to sample the new line, with public operation starting on 11 February. This was followed by the Droylsden–Ashton section on 9 October the same year.

Until the Ashton line opened, Piccadilly station undercroft was a terminus, but trams now approach the station from both ends with the Ashton line heading out at the north end of the tram station and trams bound for all other destinations using the existing exit onto London Road. On leaving Piccadilly undercroft, the line runs through a series of new medium- and high-rise residential developments in the northern part of Manchester city centre. Historically part of Ancoats, this area is nowadays known as New Islington from which the first stop gets its name. Most of the first few miles of the route are on reserved track apart from a short section of on-street running on Merrill Street between New Islington and the next stop, Holt Town, after which the River Medlock runs to the right of the line and then passes beneath it. The line then passes under the Ashburys West Junction–Philips Park Jn/Baguley Fold Jn line, which is now used only by freight and empty coaching stock trains and occasional

Above & above right: Two views of M5000 car 3038 waiting at the then temporary terminus at Droylsden with a Bury service on 23 February 2013. Work on the rest of the route to Ashton-under-Lyne is underway just beyond the end of the platform.

diversions, before reaching Etihad Campus. This stop serves the adjacent Sportcity complex, which houses the City of Manchester Stadium built for the 2002 Commonwealth Games and now home of Manchester City Football Club. A notable feature of Etihad Campus stop is that it has staggered platforms either side of Joe Mercer Way, the pedestrian bridge over the line and the Ashton Canal.

After Etihad Campus, the line take a sharp right turn as it runs through a short tunnel under the A6010 Alan Turing Way, which we then run alongside as far as the junction with the A662 Ashton New Road. Here the line curves sharply to the left just before Velopark stop, which serves the Manchester Velodrome cycle racing track and the Academy Stadium, both of which also form part of the Etihad Campus. The line then runs alongside the A662 until it reaches another bridge across the Ashton Canal, where it turns right and crosses over

the Ashton New Road and then over Croft Street and Clayton Lane. St Cross church, Clayton, is on our left as we arrive at Clayton Hall stop, where the reserved track gives way to on-street running on the A662 for most of the journey from here onwards. Much of the route is dominated by houses, small shops, schools and open spaces, with high density terraced housing gradually giving way to lower density housing as we get further away from central Manchester.

There are further stops at Edge Lane, Cemetery Road, Droylsden, Audenshaw, Ashton Moss and Ashton West. At Cemetery Road stop the track swings off the road onto a segregated area, then back onto the road immediately after leaving the stop. Audenshaw stop is likewise located on a segregated section, and after leaving the stop the line crosses over the road and runs on a dedicated right of way to the left of the A662. By this time we are passing through residential areas on the left and retail

M5000 car 3055 passes some old mill buildings as it approaches Piccadilly with a Droylsden–Bury service on 25 May 2013. *Robert Pritchard (3)*

Above: M5000 car 3052 approaches New Islington with a Droylsden service on 25 May 2013. *Robert Pritchard*

and industrial units on the right. At Audenshaw the A662 becomes the A635 Manchester Road, and the line then runs past the back of the Snipe pub and alongside the A6140 Lord Sheldon Way, which now heads in a north-easterly direction. This road crosses over the Manchester Outer Ring Road on a viaduct, and the line follows the left-hand side of the roundabout and then runs down the central reservation of the A6140 almost all the way to Ashton-under-Lyne. Shortly before arriving at the terminus, the line crosses over to the right-hand side

Above: M5000 car 3047 calls at Etihad Campus with a Droylsden service on 19 April 2013. *Alan Yearsley*

of the road, then curves to the right. A mini retail park housing a Sainsbury's and a Marks & Spencer store can be seen on the left at this point, and this is followed by the Manchester IKEA store immediately before we arrive at Ashton, one of five island platform stops on the East Manchester Line (the others

being Edge Lane, Droylsden, Audenshaw and Ashton Moss). Just beyond the tram terminus is Ashton-under-Lyne bus station, which was being rebuilt into a combined bus and tram interchange at the time of writing. The heavy rail station, on the Manchester–Huddersfield line, is a short walk further along the A6043 and then down Turner Lane to the left.

Above: Another view of Ashton terminus also taken on the first day of passenger services, this time looking in the other direction and with M5000 car 3023 awaiting departure with a Bury service.

Right: An aerial view of the approach to the Manchester Airport terminus.
Paul Abell (2)

Below: Construction work in progress on Poundswick Lane, Wythenshawe on 7 February 2014.
Robert Pritchard

MANCHESTER AIRPORT

Work to build the Manchester Airport line started in early 2011 and the line opened on 3 November 2014, over a year ahead of schedule. The Airport line diverges from the East Didsbury line just south of St Werburgh's Road stop, then takes a sharp right curve over Chorlton Brook before crossing over and running down the central reservation of Mauldeth Road West with Chorlton Park alongside us as we approach the first stop, Barlow Moor Road, located at right angles to its namesake road (on which part of the original Manchester Corporation tram network ran). Mauldeth Road West then becomes Hardy Lane, and trams share road space with other traffic.

After leaving Hardy Lane the line runs through an area of open space with Chorlton-cum-Hardy Golf Club on the left and Chorlton Ees & Ivy Green Nature Reserve on the right. We cross over the River Mersey and run along a low embankment near to Sale Water Park, which includes a 52 acre (21 hectare) artificial lake and a water sport centre. After calling at its namesake stop, the line curves to the left and runs alongside the M60 Manchester Outer Ring Road in a south-easterly direction as far as the end of Fairy Lane, then crosses over the M60 on a viaduct, which takes us to the next stop, Northern Moor.

The line then continues along a dedicated right of way originally earmarked for a new road scheme behind the back gardens of houses as far as the crossing with Kerscott Road, then takes a sharp left

Above: Amid glorious autumn colours M5000 car 3082 leaves Sale Water Park stop as it heads away from the camera with a Manchester Airport–Firswood service on 31 October 2016. *Robert Pritchard*

Right: M5000 car 3068 departs from Shadowmoss with a Cornbrook–Manchester Airport service on 5 November 2014, two days after the start of services to the airport. On the right car 3074 arrives with a Manchester Airport–Cornbrook service. *Steve Hyde*

Below: M5000 car 3067 departs Northern Moor on a driver training run on a wet 26 September 2014. *Steve Hyde*

turn immediately followed by a sharp right turn to take us to Wythenshawe Park stop. We then cross over the B5167 Wythenshawe Road and run on-street on Moor Road, from which the next stop takes its name. Here the line swings off the road onto a short reserved track section, then runs down the outer lane at either side of Southmoor Road as far as Royal Oak Road. To accommodate the tramway the junctions of Moor and Southmoor Roads with Altrincham Road had to be remodelled and Moor Road south from

Below: M5000 car 3107 crosses the River Mersey between Barlow Moor Road and Sale Water Park stops with a Manchester Airport service on 31 October 2016. *Robert Pritchard*

Left: M5000 car 3033 has just left Shadowmoss and is running alongside the A555 Ringway road with a Manchester Airport service on 12 November 2017. Its destination has already been changed to Deansgate-Castlefield for the return journey.

Below: M5000 car 3033 is seen on the long straight stretch of track between Manchester Airport and Shadowmoss with a Manchester Airport–Deansgate-Castlefield service on 12 November 2017.

Parklands Road and Southmoor Road to Royal Oak Road were widened.

At the junction of Royal Oak Road, the line reverts to reserved track on the left-hand side of Southmoor Road. A new bridge alongside Southmoor Road carries the line over the Stockport–Altrincham heavy rail line used by Manchester–Altrincham–Chester DMUs and by freight trains, and just beyond this bridge is Baguley tram stop. The line then continues along the same alignment until it reaches the next stop, Roundthorn, after which it turns left onto Hollyhedge Road, still running on reserved track to the left of the lanes used by road vehicles. Martinscroft stop is located adjacent to St Martin's parish church, which

Above: M5000 car 3108 has just left Manchester Airport and passed underneath the A555 with a Deansgate-Castlefield service on 12 November 2017.
Robert Pritchard (3)

M5000 car 3052 approaches Manchester Airport on 12 November 2017. In the background car 3091 heads back to Deansgate-Castlefield. *Robert Pritchard*

M5000 car 3063 arrives at Barlow Moor Road bound for Manchester Airport in the early morning of 3 November 2014, the first day of public operation on the airport line. *Steve Hyde*

M5000 car 3080 arrives at Manchester Airport with a service from Cornbrook (already showing Cornbrook on the destination display ready for its return journey) on 8 November 2014. The other platform is rarely used and is there in readiness for a future extension to Terminal 2 and hopefully the western loop. *Robert Pritchard*

can be seen on the left. The line then continues in a south-easterly direction, crossing over the M56 motorway just under halfway between Martinscroft and the junction with Brownley Road, where after running on-street along Hollyhedge Road it turns right and heads due south. Just after this we reach the next stop, Benchill, and the line then continues in the same direction, running at the right-hand side of the road as far as Crossacres stop.

The line then turns right by Wythenshawe fire station and runs alongside Poundswick Lane to Wythenshawe Town Centre stop, located adjacent to Wythenshawe Interchange, a new bus station opened in July 2015, and the town's main shopping area known as Wythenshawe Civic Centre. From here the line runs along Ainley Road and Fleming Road, then turns left onto Simonsway. The next stop, Robinswood Road, has staggered platforms, with the southbound platform being located just before and the northbound platform just after the junction with Brownley Road and Ruddpark Road. After this the line, still running alongside Simonsway, curves to the right as it heads towards the penultimate

intermediate stop at Peel Hall, located opposite the end of Peel Hall Road, which gets its name from an Elizabethan moated manor house demolished in the 1960s. After Peel Hall, the line turns right onto Shadowmoss Road and runs alongside Atlas Business Park on the left, which continues as far as Shadowmoss, the last stop before the airport. We then pass one of the airport's main long-stay car parks on the left, then turn right and run alongside the A555 Ringway Road, with the village of Ringway (after which Manchester Airport was originally named) being located on our right. The line passes through a short underpass beneath the A555, then the heavy rail line trails in on the left with trams and trains terminating alongside each other at the airport station. There are two Metrolink platforms; however, at present the platform closest to the heavy rail part of the station is rarely used and is there in readiness for a future extension to Terminal 2 and hopefully the western loop (see Chapter 8 on future plans).

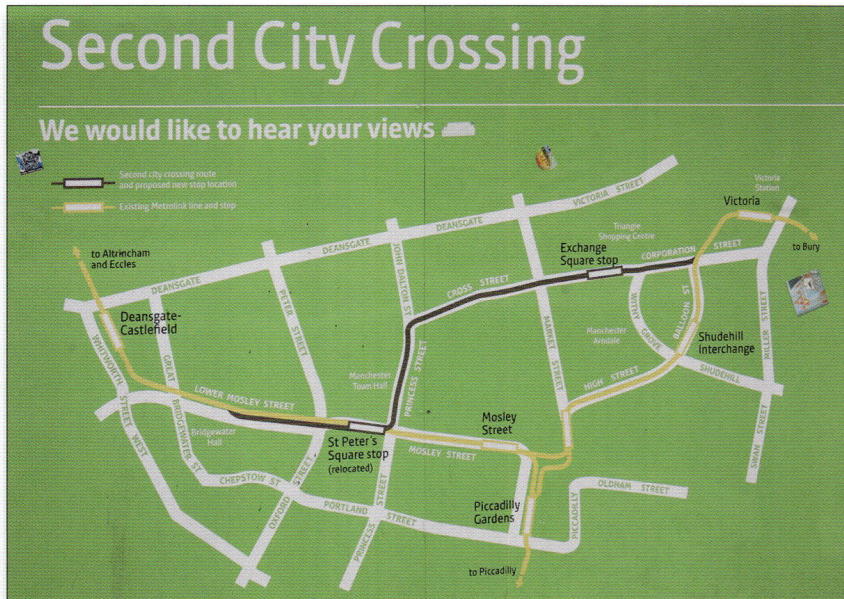

Above: A map showing the route of the Second City Crossing outside Victoria station on 16 February 2012, before the start of construction work on 2CC. The 2CC route is shown in grey and the original Metrolink route through the city centre in yellow.

Right: Construction work in progress at St Peter's Square in preparation for the Second City Crossing on 18 September 2014, including relocating the cenotaph which can be seen on the right. In the middle of this photo is Manchester Town Hall extension and, on the left, the Midland Hotel. The main Town Hall building is just visible on the right.

SECOND CITY CROSSING

The second newest part of the Metrolink network is the Second City Crossing (also referred to as 2CC), a new line through Manchester city centre first proposed in 2011 as a way of enhancing capacity and to relieve congestion on the existing route via Mosley Street with

the opening of the Phase 3a and 3b extensions. Following a public inquiry the 2CC route was approved by Transport Secretary Patrick McLoughlin and then signed off by the Greater Manchester Combined Authority in October 2013. Utilities diversion work then started in January 2014, and construction work then began on Exchange Square tram stop with the first tracks of the new line being laid in November 2014. St Peter's Square had to be remodelled and the cenotaph resited to accommodate the expanded tram stop. The northern section of 2CC between Victoria and Exchange Square opened on 6 December 2015, allowing a

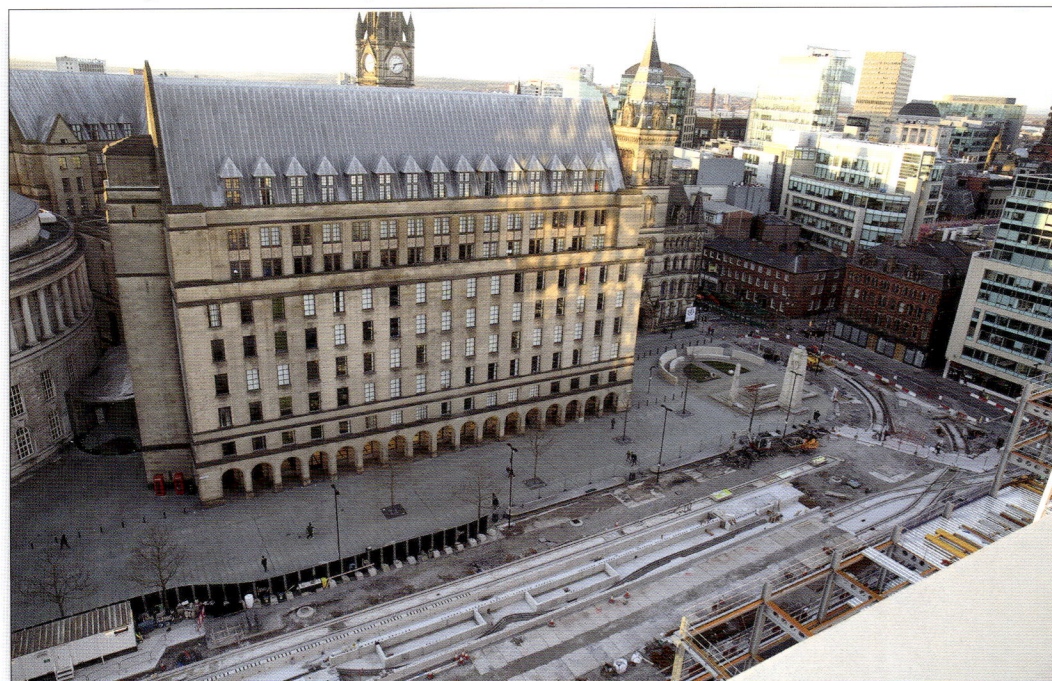

Left: An aerial view of work on the new platforms at St Peter's Square on 30 March 2016, with the Town Hall extension adjacent to the tram stop. Manchester Central Library is just visible on its left, and on its right the main Town Hall building. *Robert Pritchard (3)*

Above: M5000 car 3088 arrives at Exchange Square with a Shaw & Crompton service on 16 February 2016. The tall structure visible above the rooftops on the right is the Urbis building, home of the National Football Museum since 2012. *Robert Pritchard*

Above: M5000 cars 3079 and 3074 are seen on Cross Street on the Second City Crossing approaching Albert Square with an East Didsbury service on 14 June 2019. *Alan Yearsley*

Shaw & Crompton–Exchange Square service to operate. A tram ran on test along the entire 2CC route for the first time on 1 December 2016, with the route fully opening on 26 February 2017.

2CC diverges from the existing city centre route at Victoria station, turning right along Corporation Street with the Co-operative Bank headquarters on the left and Cathedral Gardens and the National Football Museum on the right. This is followed by the Corn Exchange and its many bars and restaurants on the right and the Arndale Centre on the left as the 2CC line reaches its only intermediate stop at Exchange Square. As Corporation Street becomes Cross Street the line continues past the Royal Exchange on the right, and at least at the time of writing there was an impressive looking mural on the side of the Arndale Centre opposite the theatre bearing the words "The Arndale: The city's centre for 40 years" and featuring two M5000 trams. After crossing over King Street, the line reaches Albert Square, then turns left along Princess Street and runs alongside Manchester Town Hall on the right before reaching St Peter's Square, where it turns right and rejoins the existing network.

TRAFFORD PARK

Metrolink's newest route is the Trafford Park line. As mentioned in passing in the previous chapter, a line to Dumplington was first suggested in 1984 by David Holt, the author of Platform 5 Publishing's 1992 Metrolink book, as a possible future extension to serve the expanding Trafford Park industrial estate when Metrolink was being planned in the 1980s. This eventually evolved into the Trafford Park extension, for which statutory powers were granted under the Greater Manchester (Light Rapid Transit System) Act 1992 and the Greater Manchester (Light Rapid Transit System) (Trafford Park) Order 2001. The then Transport Secretary Chris Grayling granted a Transport & Works Act Order (TWAO) for the 3.4 mile (5.5 km) line in October 2016 and construction work started in January 2017. The line opened to passengers on 22 March 2020, eight months earlier than originally planned, albeit without ceremony because of the Covid-19 pandemic (and despite suggestions that its opening might be delayed for this reason). Its first day of public operation was also the penultimate day that non-essential journeys by public transport were allowed in England until July, as a lockdown was imposed from the evening of 23 March with only essential journeys by key workers being permitted.

Above: On 26 February 2017, the first day of public service along the entire length of the Second City Crossing, M5000 car 3095 passes Manchester Town Hall and Albert Square as it heads towards Exchange Square with a Rochdale service, while on the left 3084 makes for St Peter's Square with an East Didsbury service.

When TfGM was applying for powers to build the line, Granada TV was concerned that the route would run close to the Coronation Street set at Salford Quays, which ITV believed could lead to problems with vibrations, "squealing wheels" from a 90 degree curve, and noise from loudspeaker announcements at the nearby tram stop when episodes of the UK's longest-running soap opera are being filmed. ITV said it did not know about the proposed tram extension when it built the Coronation Street set at this location in 2013, and requested that an alternative route be approved. However, Mr Grayling claimed that a route bypassing the quay front at MediaCityUK "would not offer the same level of benefit". Curve lubricating equipment has been installed to address the potential noise issue.

Manchester United Football Club was also concerned that the new road layout in the area served by the extension would create danger for the 75 000 plus supporters leaving the United ground on match days. Because of this, Mr Grayling granted the Transport & Works Act Order on condition that measures be put in place to enable traffic flow to United FC's Old Trafford stadium and to limit the impact of fans travelling to and from the ground including the closure of Wharfside tram

Left: Construction work in progress at Albert Square in preparation for the Second City Crossing on 16 February 2016, against a backdrop of Manchester's magnificent Town Hall. *Robert Pritchard (2)*

Above: On 22 March 2020, the day before the country entered the first lockdown period of the Covid-19 pandemic, TfGM opened the Trafford Park line following a decision that it would provide useful additional socially distanced capacity into Trafford Park. Here car 3073, suitably decked out with advertising vinyls, stands at intu Trafford Centre stop waiting to depart on the first public service inbound to Cornbrook. *Steve Hyde*

Above: The intu Trafford Centre terminus, with provision for a proposed extension to Port Salford. *Paul Abell*

stop on match days until there is a proven method of managing traffic before and after matches. The proposed alignment was originally intended to run along Trafford Wharf Road but was diverted onto the canalside following concerns about possible interference with instrumentation at Kratos Analytical, a biochemical and surface analysis instrument manufacturing company based in the area.

Running mainly on dedicated rights of way or reserved track, the branch diverges from the Eccles line at Pomona, continuing straight on and then curving to the right as it passes under Trafford Road and runs alongside the Manchester Ship Canal. Wharfside, the first intermediate stop, is reached just after the Trafford Road overbridge, and the line then runs parallel to Trafford Wharf Road on the left and opposite Salford Quays on the right as far as the next stop, Imperial War Museum, located adjacent to the Imperial War Museum North (which can also be reached via a footbridge over the Ship Canal from the Lowry Centre, a short walk from MediaCityUK, making the new line a quicker way of reaching the latter location). Also on the right are the ITV studios where the Coronation Street set is located, and the line then takes a sharp left turn along Warren Bruce Road before curving to the right at Village Circle roundabout. Here the line reaches its next stop, Village, then continues along Village way to Parkway park and ride stop, crosses over the Parkway Circle roundabout and curves to the left along the median strip of Park Way before entering a segregated alignment taking it over a new bridge across the Bridgewater Canal. At the end of this section, adjacent to the Coppice Hall Farm carvery, the line takes another right turn and runs alongside Barton Dock Road for the last leg of the journey. There is one more intermediate stop at Barton Dock Road, where a new £250 million indoor tropical spa is proposed as a replacement for the EventCity complex. The Trafford Centre terminus is adjacent to the Trafford Centre, a large shopping and leisure complex opened in 1998 which, just before the inauguration of the new line, reported financial difficulties including a £2 billion loss in one year.

Below: M5000 car 3064 departs from intu Trafford Centre with a Cornbrook shuttle on 23 March 2020, the second day (and first weekday) of public service on the Trafford Park line. In the background car 3067 can be seen in the other platform at the new terminus. *Paul Abell*

M5000 car 3068 approaches Wharfside with an intu Trafford Centre–Cornbrook shuttle, also on 23 March 2020. On the other side of the Manchester Ship Canal can be seen the Lowry Outlet Mall shopping centre. *Paul Abell*

M5000 cars 3082 and 3096 are seen running alongside the old Pomona Docks wharf on the alignment of the former Manchester Ship Canal Railway as they approach Pomona on a signalling test run on the Trafford Park line on 17 November 2019. *Steve Hyde*

CHAPTER 6:

THE TRAM FLEET, DEPOTS
& SIGNALLING

Metrolink's original fleet consisted of 26 T68 trams numbered 1001–1026, supplied by GEC-Alsthom and built by the Firema Consortium in Milan, Italy. The T68 was in many ways similar to designs that Firema had recently supplied to the Italian cities of Genoa, Naples and Rome. Unlike all other new UK tram networks, Metrolink opted for high-floor vehicles to avoid the need to raise the track or lower the platform height at former heavy rail stations and because low-floor tram technology was in its infancy when the network was at the planning stage.

Construction of the T68s was split between four Firema plants: eight of them were built at Casaralta (Bologna), seven at Fiore (Caserta, north of Naples), seven at Officine Casertane (also at Caserta) and four at Officine di Cittadella (Padua). Electrical equipment supplied by GEC-Alsthom at Preston and Trafford Park, and air equipment supplied by Davies & Metcalfe of Romiley, was shipped to Italy and fitted to the trams. Design work was carried out jointly by GEC-Alsthom at Trafford Park and by Firema in Milan. The first tram, 1001, was shipped by road and ferry across Europe from Italy to the UK in August 1991, arriving at Queens Road depot on 29 August. 1001 was delivered complete, but the next tram to arrive, 1002, was delivered to Queens Road on 28 September with its two sections and three bogies arriving separately. Problems with the organisation of such a long load across multiple borders led to 1002 and the remaining 24 T68s being shipped in sections. The third vehicle arrived on 16 October, after which one tram was delivered per week, each taking three to four days to complete its journey across Europe, with deliveries being completed by February 1992.

TECHNICAL AND INTERIOR DETAILS

The T68s were two-section 29 m articulated cars with four GEC 130 kW traction motors. The two halves were almost identical, with one section having a pantograph at the outer end and being designated as section A, the other being section B. Scharfenberg couplers were provided to enable the trams to run in pairs when required. At least five separate braking systems were fitted: an air-applied spring-released disc service brake, a spring-applied air-released disc parking brake, a regenerative dynamic service brake, a rheostatic dynamic service brake, and battery-powered magnetic track emergency brakes. Each tram had three bogies, with the two bogies at the outer ends being powered and the unmotored centre bogie mounted beneath the articulation.

There were four pairs of externally hung sliding doors (two on each section): one at the outer end and one close to the articulation. In the early days of Metrolink a number of city centre tram stops had dual height platforms, so the T68s had retractable steps for ease of boarding and alighting to and from the lower part of the platform. The trams would normally be positioned at the platform such that the doors at the inner end of each section would be located adjacent to the raised section of the platform. Each pair of doors was equipped with a door open button on the outside of the door itself and alongside the doors inside the vehicle, with a red surround that illuminated to indicate that the doors could be opened. On arrival at each stop, a single "ping" would sound to indicate that the doors had been released, followed by a "beep beep beep beep" warning signal before the doors were closed. The T68s had a whistle

Above: T68 car 1002 being delivered to Queens Road depot in September 1991. Apart from car 1001 all T68s were delivered in sections, and the gantry arrangement visible on the right was used to lift the two sections of the vehicle to enable them to be linked together. *Steve Hyde*

Left: Pioneer T68 car 1001 stands outside Queens Road depot in the autumn of 1991. *Peter Fox*

Below: T68 car 1014 tows T68A car 2001 past Exchange Quay on the first test run on the Eccles line in July 1999. 2001 had made the first run to Broadway under power but was then towed to Salford Quays by 1014 because of Railtrack's concerns about signalling interference around Cornbrook. It was rare to see a T68 coupled to a T68A. At this time car 1014 still carried the original version of Metrolink livery with the doors in the same two-tone grey as the rest of the tram. *Steve Hyde*

derived from an LMS steam locomotive whistle adapted for air operation and their present day successors, the M5000s, have an electronic whistle, a familiar though unconventional sound in Manchester city centre.

Each T68 had a total of 84 seats, of which two were pull-downs located adjacent to the doors at the inner end of section A. Section B had a dedicated wheelchair space in place of the pull-down seats. Apart from the pull-downs all seats were arranged unidirectionally, almost all in a 2+2 layout except for those adjacent to the articulation which were 2+1 and the four seats in the articulation itself, bracketed from the articulation arch, which were 1+1.

In 1995 car 1022 was experimentally fitted with a longitudinal seating layout in the centre section of each saloon in an attempt to increase overall capacity. A different design of seat was used, with slightly higher backs than the standard T68 tram seats. This trial was not a success, and when Serco took over the Metrolink contract in 1997 GMPTE requested that the interior of this vehicle be restored to the original layout. The non-standard seats from car 1022 were then reused in the now preserved car 1007, albeit in the standard unidirectional layout.

THE T68AS

In 1999 a further batch of six T68 cars, known as T68A and numbered 2001–2006, was delivered for the Eccles line. In most respects the T68As resembled the standard T68s but had digital destination displays (instead of the time-honoured destination blinds), modified controls, and covered bogies to make them more suited to the high proportion of street running on the Eccles line. The T68As also had retractable couplers (despite the fact that they only ever worked singly unlike the standard T68s) and had only 82 seats instead of 84, with a second wheelchair space in place of the two pull-down seats.

Left: T68 car 1015 leads car 1003 out of Piccadilly station undercroft towards Piccadilly Gardens with a Bury service in spring 1995, before the rebuilding of Piccadilly station for the 2002 Commonwealth Games.

Below: T68 car 1022 arrives at Piccadilly Gardens with a Piccadilly-bound service in spring 1995, before the redevelopment of the Gardens in the early 2000s. *Peter Fox (2)*

NAMES

All of the T68s (including mock-up car 1000) and T68As carried names for at least part of their lives. These were applied as vinyls on a black, red or turquoise background adjacent to the cab front, although some of the names were short-lived (names carried for only one day have been omitted) and some of them latterly became so faded as to be illegible. Most of the names were after famous Manchester personalities, places, achievements or company sponsorship (in the latter case the names were basically a form of advertising):

1000: The Larry Sullivan
1001: Children's Hospital Appeal I/ System One/Da Vinci
1002: Manchester Arndale Voyager/The Mary Sumner/Virgin Megastores
1003: The Robert Owen/Children's Hospital Appeal II/Vans. The Original since 1966
1004: The Robert Owen/Vans. The Original since 1966
1005: The Greater Altrincham Enterprise/The Railway Mission
1006: Vans. The Original since 1966
1007: The Guinness Record Breaker/Air Malta/Sony Centre Arndale/East Lancashire Railway
1008: The Manchester Airport/Erotica G-Mex 2004/Steve Hyde
1009: CIS 125 Special/The Co-operative Insurance/ Virgin Megastores
1010: Manchester Champion/West One
1011: Sponsored by Tesco/Superb/Virgin Megastores/System One/ Vans. The Original since 1966
1012: Kerry/Catherine Hallett/Virgin Megastores
1013: The Fusilier/The Grenadier Guardsman
1014: Manchester 2000/The City of Drama/Margaret Richardson/ Christies/The Great Manchester Runner/Vans. The Original since 1966
1015: Sparky/Magic 1152/Skill City/Burma Star
1016: Signal Express/Erotica G-Mex 2004/Virgin Megastores/T68 farewell* (*applied for T68 farewell tour in 2014)
1017: Rosie/Bury Hospice

1018: The Hire Flyer/Sir Matt Busby/Waterstones Manchester Arndale Now Open/Electra
1019: The Eric Black
1020: The David Graham C.B.E./Lancashire Fusilier/Mary Poppins
1021: The Greater Manchester Radio/Sony Centre Arndale/ Starlight Express
1022: The Manchester Evening News/The Graham Ashworth/ Poppy Appeal* (*Name applied twice, as The Poppy Appeal the first time and Poppy Appeal second time)
1023: Mike Mabey (name applied after withdrawal)
1024: The John Greenwood
1025: The Christie Metro Challenger/Fred G Fitter
1026: The Power
2001: The Joe Clarke OBE/West One
2002: Sony Centre Arndale
2003: Traveller 2000/Dave Hansford
2004: Salford Lads' Club
2005: WHSmith West One
2006: City of Salford 2000/West One/Sony Centre Arndale

Two more views of car 1022 on a Bury service between Piccadilly Gardens and Victoria, also taken in spring 1995.

Above: 1022 calls at the old northbound Market Street stop (which was replaced by the present day island platform at Market Street in 1998, served by trams in both directions).

Right: 1022 heads away from the camera as it passes the former southbound High Street stop. *Peter Fox (2)*

Left: T68A car 2003 departs from Broadway with an Eccles–Victoria service on 8 October 2008. Note that the destination display is not working so the tram has a printed destination on the driver's dashboard instead. *Robert Pritchard*

Above: T68A car 2001 passes Manchester Central Library (left) and the Town Hall extension (right) as it arrives at St Peter's Square on 5 October 2013. An unidentified M5000 car is just visible on the right. *Andrew Thompson*

Above: T68A car 2003 on the route for which the T68As were originally built, at the terminus at Eccles on 3 May 2010. *Robert Pritchard*

Above: In 2004 T68 car 1015 was outshopped in a special livery to publicise GMPTE's "Get Our Metrolink Back on Track" campaign to save the Phase 3 extension schemes from being abandoned. This tram is seen approaching St Peter's Square on an Altrincham–Bury service on 10 September that year. *Alan Sherratt*

Left: In 1995 T68 car 1022 received a modified interior layout with longitudinal seating in the centre section of each saloon in an attempt to increase overall capacity. This trial was not a success and within two years this vehicle reverted to its original layout with unidirectional seating throughout. *Peter Fox*

Right: 1016 and 1007 pause at MediaCityUK on the T68 farewell tour on 26 May 2014. Car 1016 (nearest the camera) appropriately carried the name "Farewell T68". Both trams had been out of service for some time and were specially prepared for the tour. They were accompanied by technicians throughout the trip but both vehicles behaved impeccably. *Steve Hyde*

LIVERIES AND REFURBISHMENT

When built, the standard T68s carried the original Metrolink livery of white with a dark grey skirt and window surround, and a turquoise strip running along the skirt beneath the doors. The T68As were delivered in a modified version of this livery with turquoise doors, and in due course this would be extended to the standard T68 fleet on refurbishment with a later variant of this livery also including a turquoise strip along the top of the body beneath the roof. When carrying the original livery, the fleet number was only displayed on each side at the front and back end of each section, but all T68s and T68As latterly also carried their numbers on the cab front with the suffix "A" or "B" to denote whether it was the "A" or "B" section. In 2004 car 1015 received a special purple and yellow livery to publicise the "Get our Metrolink Back on Track" campaign, and in late 2011 car 1003 was outshopped in Metrolink's new yellow and silver livery. At this time this vehicle also carried red bodyside adverts for the Manchester Christmas markets.

In the early 2000s three of the standard T68s (1005, 1010 and 1015) were modified with retractable couplers and covered bogies to enable them to run to Eccles. Then in 2008 the then operator of the Metrolink franchise RATP Dev announced an 18-month refurbishment programme for the T68 and T68A fleet to allow the entire fleet to run on all the lines in operation at that time and to make the vehicles fully compliant with the Disability

Above: In February 2020 T68 car 1023 was moved to Crewe Heritage Centre, where it is seen on display on 12 August 2020. At the time of going to press this tram was still at Crewe, and is expected to remain there until it can be accommodated at Heaton Park. *Cliff Beeton*

Discrimination Act and the Rail Vehicle Accessibility Regulations. These modifications were then extended to all T68s except cars 1018, 1019 and 1020. Changes were also made to the interior layout so that both the T68s and the T68As had 86 seats of which four were pull-downs. Other modifications included internal digital information displays, enhanced lighting and CCTV cameras, improved braking

Above: 1003 was the only T68 car to receive Metrolink's new yellow/grey livery. It is seen at Old Trafford on 17 April 2012 carrying advertising decals for the Imperial War Museum North on an Altrincham service. *Robert Pritchard*

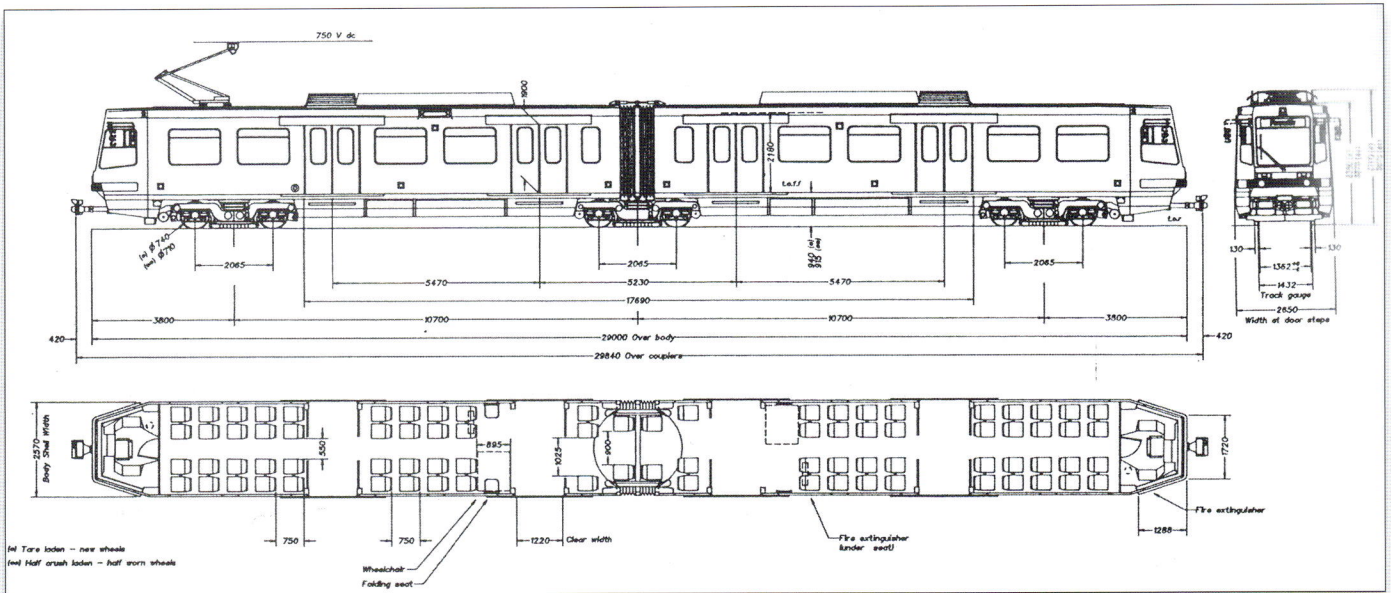

Above: A diagram showing the dimensions and interior layout of the T68 trams in as-built condition, taken from Platform 5's 1992 Metrolink book by David Holt.

systems, and automated announcements (previously the stops and the destination were announced manually by the driver).

TABLE 1:
T68/A TECHNICAL DATA (AFTER REFURBISHMENT)

Built	1991–92 by Firema, Italy (T68), 1999 by Ansaldo, Italy (T68A).
Wheel arrangement	Bo-2-Bo
Traction motors	Four GEC of 130 kW
Line voltage	750 V DC
Track gauge	1435 mm
Seats	82 (+4 pull-down)
Standing capacity	122
Weight	45 tonnes.
Braking	Rheostatic, regenerative and disk. Emergency track (T68), magnetic track (T68A).
Wheel diameter (new)	740 mm
Couplers	Scharfenberg.
Maximum speed	50 mph (80 km/h)
Doors	Externally hung sliding.
DIMENSIONS	
Length	29.0 m
Width	2.65 m
Height	3.36 m
PERFORMANCE DATA	
Acceleration	1.3 m/s^2
Deceleraton (service brake)	1.3 m/s^2
Deceleration (emergency brakes)	2.6 m/s^2

It was hoped that the refurbishment programme would extend their working lives by at least ten years, but in the event the entire T68 and T68A fleet would be superseded by the new M5000s only a few years later (see below). The last three T68s were withdrawn on 10 February 2014, followed by the last two T68As, 2001 and 2003, on 30 April with the last T68A-operated service being the 20.03 Bury–Abraham Moss worked by car 2003. Cars 1007 and 1016 then ran a farewell tour on 26 May taking in the Eccles, Bury and Altrincham lines. The T68/As were not gauged to run on most of the Phase 3 network, however, so the tour only covered the Phase 1 and 2 lines. Most of the T68 fleet was then scrapped but car 1007 was presented to the Heaton Park Tramway, this one being chosen because it was the first tram to run through Manchester city centre on the first day of passenger services between Victoria and Deansgate on 27 April 1992. 1007 was also the fleet number of Manchester Corporation Tramways' official last tram in 1949. At the time of writing it is in store at Metrolink's Trafford depot but is due to be moved to Heaton Park in due course. Car 1023 was moved from Trafford depot to Crewe Heritage Centre in February 2020 and is expected to remain

there until it can be accommodated at Heaton Park. In the same month cars 1020 and 2001 were scrapped on site at Trafford depot meaning that there are now no surviving T68As. 1003 was purchased by Greater Manchester Fire and Rescue Service for use as a training rig, and a further four T68s (1016, 1022, 1024 and 1026) are stored at Long Marston for use in development work by UK light rail industry association UK Tram.

THE M5000S

In 2007 GMPTE ordered eight Bombardier Flexity Swift M5000 trams based on the existing K5000 design in Cologne and Bonn, Germany. This was the first of several orders placed for M5000s between 2007 and 2018. The first eight vehicles, numbered 3001–3008, were intended to supplement the existing T68 and T68A fleet and to relieve overcrowding on the Bury–Altrincham line. Four more cars (3009–3012) were ordered in November 2007 for the new MediaCityUK extension. This was followed by a further order for 28 cars (3013–3040) in June 2008 for the Rochdale and St Werburgh's Road extensions. After funding had been secured for the Phase 3b extensions, another eight were ordered in March 2010 for the East Didsbury and Ashton routes followed by another 14 in July 2010 for the Manchester Airport and Rochdale lines.

By this time the M5000 trams were proving to be more reliable than the T68s and T68As. It was reported that the T68/A fleet averaged 5000 miles between breakdowns compared to 20 000 miles between breakdowns for the M5000s at the time of their introduction. Many of the T68/As were also found to be suffering from corrosion to their solebars. The T68As were prone to overheating in warm weather because of the relatively slow speeds on the Eccles line, although as more M5000s entered service and more modified T68s became available for use on the Eccles line T68As started to be used on the Altrincham and Bury lines where they proved to be more reliable when operating at higher speeds than in their original role. The M5000s are also 5.3 tonnes lighter at 39.7 tonnes compared to 45 tonnes for a T68/A, meaning that they use less energy and cause less wear and tear on the tracks.

Because of this, 12 more M5000s were ordered in September 2011 to replace the 12 T68s that were in worst condition. Another 20 M5000s were ordered in July 2012 after it was decided to withdraw the entire T68/A fleet. Ten more M5000s were ordered in January 2014 for the Trafford Park extension opened in 2020. This was followed by another 12 in July 2014 and a final four in September of that year, both to enable more double trams to operate. A further 27 M5000s were ordered in July 2018 to be delivered in 2020–21 to enable all services on the Bury–Altrincham line to be worked by pairs of trams and to allow additional double trams to operate on the

East Didsbury–Rochdale route and some on the Ashton-under-Lyne route. Thus, a total of 124 M5000s (numbered 3001–3124) have been delivered to date and 3125–3147 are on order. As this book was nearing completion, car 3121, the first of the latest order, arrived at Queens Road depot on 14 November 2020 and entered service on 23 December, and car 3122 was delivered on 12 December followed by 3123 on 6 February 2021 and 3124 on 30 March with the rest of the new batch expected to follow shortly.

Car 3001, the first M5000, arrived at Queens Road depot by road from Vienna, Austria via the Rotterdam–Hull ferry in the early hours of 13 July 2009, with the first M5000 entering service on 21 December that year. The M5000s are assembled at the Bombardier plant in Vienna, with the sub-assemblies and primary parts being manufactured at Bautzen, Germany, the underframes at Česká Lípa, Czech Republic, and the bogies at Siegen, Germany, with the electrical equipment being supplied by Vossloh-Kiepe in Düsseldorf.

Above: The old and the new order on Metrolink side by side: T68 cars 1013 and 1017 call at Trafford Bar with an Altrincham service on 19 April 2013, while M5000 car 3035 forms an Altrincham–Piccadilly service. The original street level station building is still in situ. *Alan Yearsley*

TABLE 2: M5000 DELIVERY DATES

3001:	13 July 2009	3032:	7 May 2011	3063:	26 January 2013	3094:	17 January 2015
3002:	18 September 2009	3033:	14 May 2011	3064:	16 February 2013	3095:	21 February 2015
3003:	12 September 2009	3034:	28 May 2011	3065:	9 March 2013	3096:	14 March 2015
3004:	10 October 2009	3035:	18 June 2011	3066:	23 March 2013	3097:	28 March 2015
3005:	24 October 2009	3036:	2 July 2011	3067:	20 April 2013	3098:	25 April 2015
3006:	14 November 2009	3037:	16 July 2011	3068:	11 May 2013	3099:	16 May 2015
3007:	21 November 2009	3038:	30 July 2011	3069:	22 June 2013	3100:	14 June 2015
3008:	December 2009	3039:	13 August 2011	3070:	29 June 2013	3101:	4 July 2015
3009:	23 January 2010	3040:	3 September 2011	3071:	20 July 2013	3102:	25 July 2015
3010:	6 February 2010	3041:	17 September 2011	3072:	24 August 2013	3103:	5 September 2015
3011:	20 February 2010	3042:	15 October 2011	3073:	21 September 2013	3104:	19 September 2015
3012:	17 April 2010	3043:	29 October 2011	3074:	12 October 2013	3105:	10 October 2015
3013:	1 April 2010	3044:	12 November 2011	3075:	2 November 2013	3106:	31 October 2015
3014:	30 April 2010	3045:	3 December 2011	3076:	16 November 2013	3107:	14 November 2015
3015:	8 May 2010	3046:	10 December 2011	3077:	7 December 2013	3108:	28 November 2015
3016:	22 May 2010	3047:	17 December 2011	3078:	18 January 2014	3109:	12 December 2015
3017:	12 June 2010	3048:	23 January 2012	3079:	8 February 2014	3110:	23 January 2016
3018:	26 June 2010	3049:	11 February 2012	3080:	1 March 2014	3111:	13 February 2016
3019:	10 July 2010	3050:	10 March 2012	3081:	22 March 2014	3112:	5 March 2016
3020:	24 July 2010	3051:	31 March 2012	3082:	12 April 2014	3113:	2 April 2016
3021:	7 August 2010	3052:	28 April 2012	3083:	10 May 2014	3114:	16 April 2016
3022:	4 September 2010	3053:	26 May 2012	3084:	31 May 2014	3115:	7 May 2016
3023:	18 September 2010	3054:	16 June 2012	3085:	28 June 2014	3116:	4 June 2016
3024:	30 October 2010	3055:	7 July 2012	3086:	19 July 2014	3117:	25 June 2016
3025:	11 December 2010	3056:	28 July 2012	3087:	16 August 2014	3118:	16 July 2016
3026:	29 January 2011	3057:	1 September 2012	3088:	13 September 2014	3119:	20 August 2016
3027:	19 February 2011	3058:	22 September 2012	3089:	4 October 2014	3120:	1 October 2016
3028:	5 March 2011	3059:	13 October 2012	3090:	25 October 2014	3121:	14 November 2020
3029:	19 March 2011	3060:	17 November 2012	3091:	8 November 2014	3122:	12 December 2020
3030:	2 April 2011	3061:	1 December 2012	3092:	22 November 2014	3123:	6 February 2021
3031:	16 April 2011	3062:	15 December 2012	3093:	6 December 2014	3124:	30 March 2021
						3125–3147:	to be delivered.

Below: The dimensions and interior layout of an M5000 car. The seating arrangement shown has 52 seats, as on cars 3001–3074. *Courtesy Bombardier*

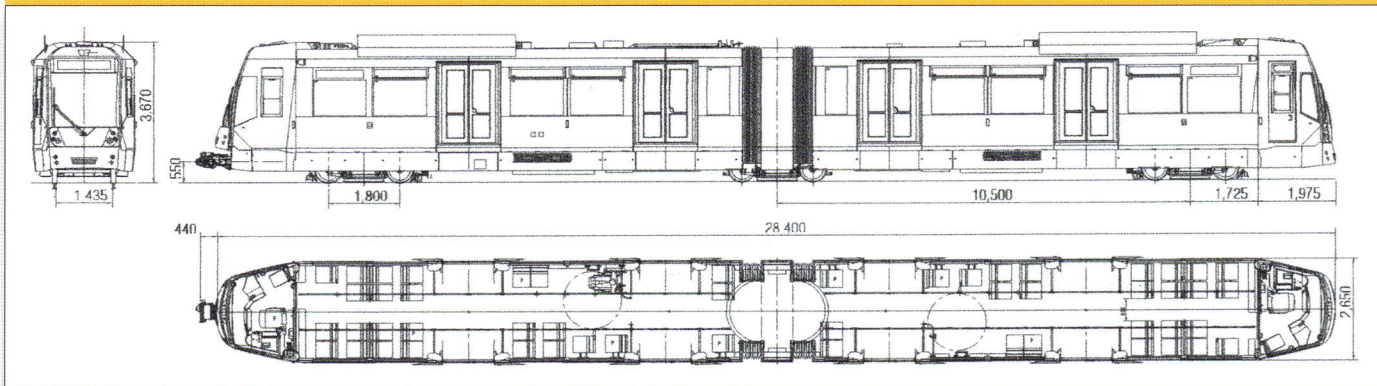

Above: A model of M5000 car 3001 on display outside the Manchester Central (formerly G-Mex) Convention Centre on 8 October 2008 during a media event at which the forthcoming arrival of the M5000 fleet was announced. (Inset: A close-up of the model of car 3001, showing the easy access between the platform and the tram.) *Robert Pritchard (2)*

As with the T68/As, the M5000s have three bogies with the outer end bogies being powered. Sliding plug doors are fitted instead of externally hung sliding doors. The M5000s are slightly shorter than the T68s (28.4 m instead of 29 m) and have a similar overall capacity to their predecessors but a much lower number of seats, giving greater standing room and more space for luggage and pushchairs etc. Although all the M5000s are largely identical in appearance, there are a few minor detail differences between the different batches. 3001–3060 are fitted with Automatic Tram Stop (ATS) and Vehicle Recognition System (VRS) and can be used on all routes. 3061–3124 (and the still to be delivered cars 3125–3147) do not have these systems and cannot be used on the Altrincham line because the Timperley–Altrincham section still has block signalling with ATS whereas the rest of the Metrolink network has Tram Management System (TMS) signalling for which all trams are equipped (see below). Also cars 3001–3074 have 52 seats and 3075–3124 have 60 (3125–3147 will also have 60). All have eight perch seats (four in each of the two wheelchair spaces). Unlike the T68/As, seating is in a mixture of longitudinal, forward and backward facing layouts. The seats immediately next to the driving cabs still offer a forward view but rather oddly the seats immediately behind those at the front end

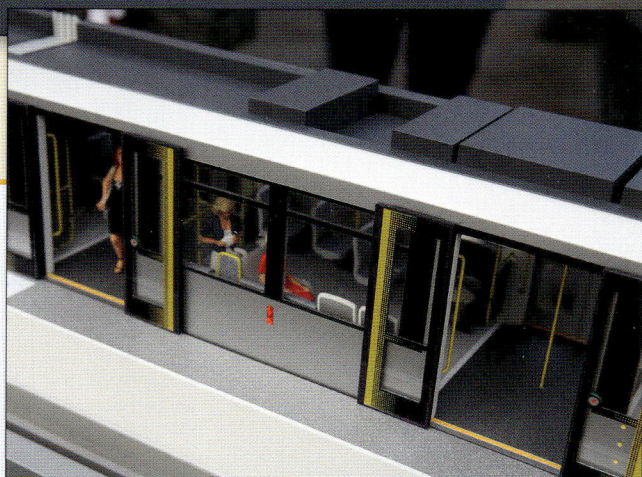

face backwards. As with the T68s in as-withdrawn condition, section A of each M5000 car has the suffix "A" after the fleet number on the cab front and section B the suffix "B". However, the M5000s have their pantographs at the inner end of section A unlike the T68s which had them at the outer end close to the driving cab.

All M5000s were delivered in Metrolink's new yellow and silver livery (which was also applied to T68 car 1003), and at the same time as the first M5000s arrived this colour scheme was also adopted for all tram stops and signage to replace the previous turquoise and grey. At the time of writing nine of the M5000s carried all-over advertising liveries:

Right: Car 3003 at St Peter's Square on the special VIP/press run to Eccles at the official launch of the first M5000 tram into service on 21 December 2009. The same vehicle returned in normal service later in the day. *Steve Hyde*

Below: M5000 car 3018 currently carries an all over advert for Phantom of the Opera. It is seen on the rear of a Bury service coupled to car 3007 at St Peter's Square on 29 February 2020. *Robert Pritchard*

3091: Go North Wales
3099: Gymshark
3103: PrettyLittleThing.com
3117: Cyperpunk 2077

In addition four M5000s have been named so far:
3009: 50th Anniversary of Coronation Street 1960–2010
3020: Lancashire Fusilier
3022: Spirit of Manchester
3098: Gracie Fields
Car 3022 also features special "Spirit of Manchester" decals inspired by the worker bee, which has become a symbol of Manchester. These were applied to honour the victims of the Manchester Arena bombing of 22 May 2017.

3018: The Phantom of the Opera
3049: BBC Tiny Happy People
3066: Clean Air Greater Manchester
3073: intu Trafford Centre
3089: Dippy on Tour

TABLE 3:
M5000 DATES OF ORDERING AND MAIN PURPOSE

BATCH	DATE ORDERED	MAIN PURPOSE
3001–3008	April 2007	Supplement T68/A fleet; Alleviate overcrowding on existing network.
3009–3012	November 2007	MediaCityUK extension.
3013–3040	June 2008	Rochdale and St Werburgh's Road lines.
3041–3048	March 2010	East Didsbury and Ashton routes.
3049–3062	July 2010	Manchester Airport and Rochdale lines.
3063–3074	September 2011	Replace the 12 T68s in worst condition.
3075–3094	July 2012	Replace the remaining T68/As.
3095–3104	January 2014	Trafford Park line.
3105–3116	July 2014	Enable more double trams to operate.
3117–3120	September 2014	Enable more double trams to operate.
3121–3147	July 2018	All Bury–Altrincham services to be double trams; additional double trams on East Didsbury–Rochdale and some on Ashton line.

TABLE 4: M5000 TECHNICAL DATA

Built	2009–20 by Bombardier, Vienna, Austria and Bautzen, Germany.
Wheel arrangement	Bo-2-Bo
Traction motors	Four Bombardier 3-phase asynchronous of 120 kW.
Line voltage	750 V DC
Track gauge	1435 mm
Seats	52 (3001–3074), 60 (3075–3147).
Standing capacity	146
Weight	39.7 tonnes.
Braking	Rheostatic, regenerative, disk and magnetic track.
Wheel diameter (new/worn)	580/510 mm
Couplers	Scharfenberg.
Maximum speed	50 mph (80 km/h)
Doors	Sliding plug.
DIMENSIONS	
Length	28.4 m
Width	2.65 m
Height	3.4 m
PERFORMANCE DATA	
Acceleration	1.08 m/s^2
Deceleraton (service brake)	1.03 m/s^2
Deceleration (emergency brakes)	2.54 m/s^2

THE M5000 INTERIORS

Above: An interior view of M5000 car 3054. The wheelchair space and emergency alarm and call button for summoning assistance is on the right.

Above: One of the outer end saloons of M5000 car 3017, showing the view of the line ahead (or, in this case, the line behind!). Rather oddly the third and fourth pairs of seats face backwards.

Above: The wheelchair space on car 3017.

Right: The driving cab of M5000 car 3023. *Robert Pritchard (5)*

Above: The inside of a pair of doors on car 3017. On the left are warnings about not boarding or alighting when the doors are closing, CCTV images being recorded and monitored, and that surfaces may be slippery when wet. On the right are notices about "touching out" when using a smartcard and the Metrolink byelaws and conditions of carriage prohibiting the consumption of food and alcohol and the conveyance of cycles (except folding cycles), dogs (except assistance dogs) and mobility scooters without a valid Metrolink Mobility Scooter Permit.

Below: For a number of years M5000 car 3091 was branded TramGB and carried a special gold livery to commemorate the achievements of the athletes in TeamGB and Paralympics GB at the 2012 London Olympics and Paralympics. Here it has just left Sale Water Park with a Manchester Airport service on 31 October 2016 (despite the destination showing Firswood). This tram has since lost this colour scheme and received an all over advert for Go North Wales.

Above: M5000 car 3089 currently carries advertising decals for the Dippy on Tour exhibition in Rochdale, although after this photo was taken the event was extended until December 2020 because of the Covid-19 pandemic. On 29 February 2020 it is seen approaching St Werburgh's Road bringing up the rear of a Rochdale service, coupled to car 3079. *Robert Pritchard (2)*

NEVER REALISED PLANS: T68 CENTRE SECTIONS, SAN FRANCISCO AND BONN TRAMS

In late 2000 the Government allocated £7 million to add a centre section to 12 T68 trams. This was intended to relieve overcrowding on the Bury–Altrincham route, but the plan was not progressed any further because GMPTE realised that it would be too expensive to obtain a sufficient level of fire safety certification for these centre sections.

Instead, in January 2002 Metrolink purchased Boeing Vertol car 1326 from the Muni Metro light rail system in San Francisco, USA. At this time these cars, which were built between 1980 and 1985, had recently been withdrawn from service. Sister vehicle 1226 was taken to the Serco Test Centre in Derby in May of that year for examination by the Railway Inspectorate to determine whether it would meet UK safety standards. A number of other members of the fleet were also earmarked for shipping to the UK for Metrolink. However, car 1326 derailed twice on one bogie during gauging trials in early February, and by mid-2002 it was decided that these vehicles would need too many modifications to enable them to run on Metrolink. 1326 was scrapped on site at Queens Road depot and 1226 languished at Derby until 2016 when it too was scrapped.

In the autumn of 2002 Metrolink considered purchasing ten B100 Stadtbahn cars of 1970s vintage from Bonn, Germany. These were then due to be replaced by K5000 cars on which Metrolink's M5000s were based. However, in the event the vehicles that had been considered for Manchester were instead purchased by the city of Dortmund in the Ruhr district.

DEPOTS

Metrolink has two depots: the original depot is located at Queens Road in Cheetham Hill on the Bury line. As well as a depot for stabling, maintenance and repairs to the tram fleet, the site, known as Metrolink House, also served as a control centre, headquarters and office space. Metrolink House was originally planned to be much larger so that it could support network expansion, but this was refused planning permission by Manchester City Council. Because of this, Metrolink had to settle for a smaller 9.9 acre (4 hectare) site with limited capacity. This led to a decision to build a second 67 000 m² depot at Old Trafford, where construction work started in 2009. At the same time Queens Road depot was expanded to increase capacity. Trafford depot was completed in 2011 and formally opened for operational staff in June 2012, and the Metrolink control room, known as the Network Management Centre (NMC), was moved from Queens Road to Trafford in May 2013. The new depot is located between Old Trafford tram stop on the Altrincham line and the north end of the East Didsbury line and has four lanes inside and 11 sidings outside the depot, giving enough space to stable up to 96 trams. Other facilities include maintenance workshops, automatic tram washing equipment, a wheel lathe, sand plant and cleaning platforms to allow access to tram interiors for preparation for service.

Routine examination, basic maintenance and repairing minor faults is carried out at Trafford. Heavy maintenance, including brake overhauls, bogie work and retyring mostly takes place at Queens Road, as there are no bogie facilities at Trafford. Installation of a wheel lathe at Trafford has enabled tyre turning to be shared between the

Above: In 2019 M5000 car 3081 sported advertising decals for IKEA's Manchester store at Ashton-under-Lyne, the destination of this tram as it heads away from the camera off Aytoun Street towards Piccadilly station on 15 June 2019. *Alan Yearsley*

Above: M5000 car 3111 carried special jigsaw decals in 2017 to mark the opening of the full length of the Second City Crossing. It is seen passing Manchester Town Hall as it approaches St Peter's Square with a Rochdale–East Didsbury service on 26 February 2017.

Above & above right: Two views of the roof of car 3047 inside Trafford depot, including a close-up view of the pantograph. *Robert Pritchard (3)*

two depots, however. Repairs to damaged vehicle bodies also take place at Queens Road.

The entire Metrolink system is monitored from the NMC. As well as controlling the signalling and tram movements, the NMC also contains a wall of CCTV monitors covering every line and tram stop. Controllers monitor the network at all times of day and night and are instantly alerted to any incidents, with each controller being responsible for a specific part of the network. Manual announcements and updates to passenger information systems at tram stops are made from the NMC, and staff also update the Metrolink Twitter feed from here. The NMC is staffed 24 hours a day 365 days a year, as it is necessary to ensure that the overhead cables are in working order even when trams are not running.

THE SPECIAL PURPOSE VEHICLE

As well as its fleet of ordinary service trams Metrolink also has a 4-wheel diesel-hydraulic multi-purpose vehicle with a crane. This is a customised version of RFS Industries' "Mantis" design of vehicle built to the order of Brown & Root Vickers, the subcontractor responsible for supplying Metrolink's maintenance equipment. It incorporates all the features of a diesel loco, mobile crane, overhead line inspection platform and a travelling workshop and store. BRV set out to identify all possible requirements and asked RFS to meet them with one robust street-compatible vehicle. This Special Purpose Vehicle (SPV) is mainly used for infrastructure maintenance work. It can also rescue failed trams but it is more common for another tram to be used for this purpose.

The SPV is based at Queens Road depot and is used, along with a small fleet of road vehicles, to access any part of the network where it is needed. It carries the number 1027 and is thus numbered in the same sequence as the now withdrawn T68 trams, the highest numbered of which was car 1026. Adhesive weight and robustness are combined in a frame welded up from rolled steel sections and plate. Its engine is a six-cylinder Caterpillar 3306B turbocharged diesel unit with an intermittent rating of 170 kW (225 hp) at 2200 rpm, similar to engines used in excavators farm machinery and other plant.

Its transmission system is a heavy duty Caterpillar power shift unit mounted integrally with the engine and comprising a single stage torque converter, planetary-type gearing and power shift clutches for forward, reverse and all speeds. Final drive is via cardan shafts to a double reduction gearbox on each of the two axles, which are fitted

Above: The bogies of M5000 car 3023, as seen from the inspection pit beneath the vehicle at Trafford depot. *Robert Pritchard*

Above: M5000 car 3047 stands in the roof inspection bay, while on the right 3006 is lifted up above one of the inspection pits at Trafford depot using Mechan jacks on 17 April 2012.

Above: This view of the sidings outside Trafford depot on 17 April 2012 shows no less than nine M5000 cars stabled. *Robert Pritchard (2)*

with rubber chevron primary axlebox suspension and monobloc wheels. The input shaft of each axledrive gearbox carries a ventilated disk acted on by the unit's straight air service brake system and air release/spring applied emergency and parking brakes. Pneumatic sanders are fitted.

There is room for up to four people in the spacious SPV cab, which can even be used by infrastructure maintenance crews to make hot drinks when working in locations not accessible to the public. The driver can sit or stand at either side of the cab to operate the controls, which are simple and can activate towed tram brakes via compatibly coded brake wire signals. The SPV can also interface with continuous air brakes on railway rolling stock via standard brake hose connections.

The RFS Mantis has a special quick-release mounting plate that enables different couplers to be fitted according to need. Metrolink's SPV has a Dellner auto-coupler at one end compatible with those fitted to the trams, and at the other end a plain coupling pocket that can be used to attach a general purpose trailer supplied by RFS for use with the SPV. The couplers on the SPV can be interchanged anywhere on the system using its hydraulic crane, which can also carry an insulated "cherry picker" platform to lift up maintenance staff working on the overhead power lines or lineside structures. The SPV can provide power for handheld worklights and air or hydraulic take-off power for appliances, plant or tools used in maintenance or recovery work. In Metrolink's case, portable electric power is provided by conventional equipment of a more portable nature. RFS can also provide a range of accessories for use with the SPV, including snowploughs, snow-blowers, and track cleaning equipment such as power vacuum suction cleaners and road sweepers.

SIGNALLING

Most of the Metrolink network uses Line of Sight signalling principles (as also used on most other UK light rail networks with street running sections) allied to a Tram Management System (TMS) supervisory system. This means that drivers must maintain a speed such that they can stop short of any obstructions simply by using the normal service brakes. Because of this, the maximum speed on the on-street sections is generally the same as for motor vehicles on the stretch of road in question (usually 30 mph in a built-up area) but in some locations trams have to observe a lower speed limit where this is necessary for reasons of track alignment. There are also sections with 40 mph or 20 mph speed limits where the line runs on or alongside highways with the same limit. On the segregated sections elsewhere the speed limit is generally 50 mph but may be lower in places as determined by TfGM to enable trams to operate safely while maintaining as short a headway between trams as possible.

However, some signalling is still necessary at road junctions and crossings, on single line sections and on the on-street sections to avoid conflict with road traffic and because trams have a degree of priority over normal road traffic at all road junctions as determined by TfGM, which has to ensure that all traffic flows as efficiently as possible at all times (including buses for which it is also responsible). Where a tram stop is located adjacent to a road junction or intersection, the tram driver presses the "ready to start" button when ready to leave the stop. This avoids the signals being set for the tram to depart too early, thereby causing delays to other trams or road traffic.

To prevent confusion with the conventional traffic signals, special white lights are used, and on the street-running sections these are

Above: Metrolink's Special Purpose Vehicle (SPV) is a customised version of RFS Engineering's "Mantis" design. It carries the number 1027 and was built by RFS to the order of Brown Root, the subcontractor responsible for supplying Metrolink's maintenance equipment. The SPV is mainly used for infrastructure maintenance work and can also be commandeered to rescue failed trams, although it is more common for another tram to be used for this purpose. *Colin J Marsden*

to locate all trams on the network within the Control Room for supervisory and performance purposes. Lineside signs are also used to give warnings or instructions to tram drivers, such as speed restrictions. Tramway signs are diamond shaped to avoid causing confusion to other road users.

All on-street signage and signals must comply with the Department for Transport Traffic Signs Manual. The only exception to this is the points indicators, which are an oddity as they are not an approved sign. This means that, in theory, each indicator installation needs its own approval. Similarly, until recently each No Entry Except Trams sign needed its own authorisation from the DfT.

TMS signalling is used on all lines except for the Timperley–Altrincham section of the Altrincham line, which still uses block signalling and conventional two-aspect red/green signals. At the time of going to press the Timperley station area, including the junctions with the centre turnback siding, was due to be equipped with TMS signalling shortly.

Block signalling was also used on the Bury line until recently, but over the last few years the route has been converted to TMS, the last section to do so being that between Whitefield and Bury Interchange, which moved over to this system on 24 August 2020.

Above: The signal at Shudehill for the northbound line towards Victoria, Bury and Rochdale is on "Stop" here as M5000 cars 3012 and 3036 approach with a Bury–Altrincham service on 18 April 2014. At this time rebuilding work was taking place at Victoria station so a single track section was in operation between Shudehill and Victoria, hence the temporary tram gate at the end of the northbound platform.
Inset: a close-up view of the signal.

often located above the traffic lights. These signals consist of five white lights arranged either horizontally, telling a tram to stop, or vertically to allow the tram to proceed. A diagonal line pointing south-east to north-west indicates that a tram may proceed if turning left, and a diagonal line pointing south-west to north-east means that a tram may proceed if turning right. Five white lights arranged in the shape of a cross (known as a "cluster") in the centre of the signal indicator has the same meaning as an amber road signal: stop if it is safe to do so.

There are also points indicators on the approach to facing points and some trailing points to inform drivers which way the points are set and locked. As with the ordinary signals, these can show either a horizontal line for stop or a diagonal dogleg shaped line to indicate that the points are set for either a left or a right turn. The points are called by a data message sent by the on-board TMS equipment on the tram. This is generated from the on-board database of topography and timetables held in the TMS system. It is communicated to the points system via a series of trackside loops and wi-fi receivers. The same series of loops and wi-fi equipment is used to send requests to traffic signal controllers, and in segregated areas TMS local controllers, for the purpose of authorising the trams' routing through the network. It is also used

Above: A busy scene at Victoria on 9 October 2019 as M5000 cars 3034 and 3039 approach with a Bury service and on the left car 3003 heads towards the city centre with a Piccadilly service. The signal next to car 3003 is showing that a left-turning tram may proceed. On the right another tram can be seen waiting to enter Victoria station from the Second City Crossing.
Inset: a close-up view of the signal. *Robert Pritchard (2)*

METROLINK SIGNALS & INDICATORS

FIXED SIGNALS

Signals normally display five white lights which are distinctive from those of the standard road traffic lights or railway lineside signals and have the meanings indicated:

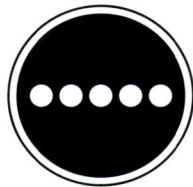

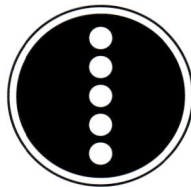

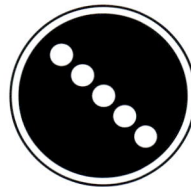

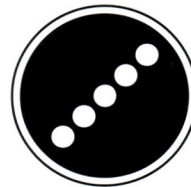

STOP	STRAIGHT THROUGH TRAM OR TRAIN MAY PROCEED	LEFT TURNING TRAM OR TRAIN MAY PROCEED	RIGHT TURNING TRAM OR TRAIN MAY PROCEED	THIS IS EQUIVALENT TO AN AMBER ROAD TRAFFIC LIGHT SIGNAL

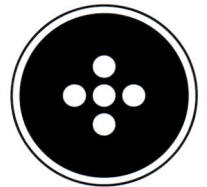

POINTS INDICATORS

Points indicators are provided at junctions to indicate the route which is set through points. At junctions where the tram and train movements can conflict with road traffic, fixed signals are provided in addition to points indicators. The aspects displayed by the points indicators are:

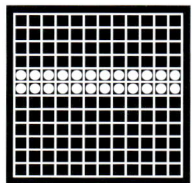

A points indicator may be passed only if it displays the correct route indication for the tram or train concerned and, where fixed signals are provided, if both points indicators and fixed signals are set for the correct route.

STOP (POINTS MISALIGNED/NOT DETECTED)	POINTS SET FOR LEFT TURNING TRACK	POINTS SET FOR RIGHT TURNING TRACK

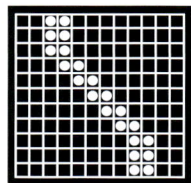

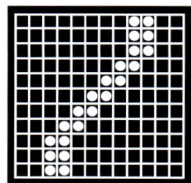

METROLINK SIGNS

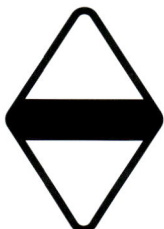

STOP AND PROCEED WHEN SAFE TO DO SO	GIVE WAY TO ANOTHER TRAM, TRAIN OR ROAD VEHICLE	MAXIMUM PERMITTED SPEED IN MILES PER HOUR	SOUND AUDIBLE WARNING	TEMPORARY SPEED RESTRICTION	TERMINATION OF TEMPORARY SPEED RESTRICTION	INSTRUCTION SIGN OBSERVE SPECIFIC INSTRUCTION ON PLATE BELOW

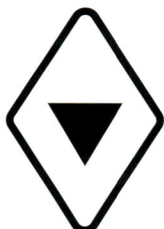

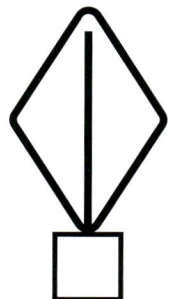

EXAMPLES OF SPECIFIC INSTRUCTIONS

S	PREPARE TO STOP AT COMMENCEMENT POINT OF INTERLOCKING AREA	**SS**	TERMINATION OF INTERLOCKING AREA
SS	INTERLOCKING AREA COMMENCES ("SAFETY SIGNALLING")	**I**	SECTION GAP
		LOS	LIMIT OF SHUNT

CHAPTER 7:
SERVICES AND TICKETING

Before Metrolink opened in 1992, GMPTE's original plans envisaged a 10-minute service frequency on the Bury–Piccadilly and Altrincham–Piccadilly routes from 06.00 to midnight, Monday–Saturday. Metrolink's original operator, Greater Manchester Metro Ltd, argued for adjustments to this pattern in line with vehicle running times and passenger demand. Because of this a 12-minute frequency was adopted, doubling to a 6-minute service in peak periods.

The current service consists of nine routes, each of which operates a 12-minute frequency throughout most of the day giving a 6-minute service on most of the network and an even higher frequency on sections served by several routes, particularly in the city centre. Most routes operate from around 05.30 to midnight (05.30 to 01.00 on Fridays and Saturdays, 07.00 to midnight on Sundays). Since 2016 there has also been a very early morning service on the Airport line aimed at airport workers and passengers with early flights. The first tram runs empty from Trafford depot to Firswood, where it leaves at 03.00 and arrives at the Airport at 03.39. Trams then run every 20 minutes from Deansgate-Castlefield to the Airport until 06.00 when the normal 12-minute frequency starts. Additional trams operate to cater for football matches at Old Trafford (Manchester United) or the Etihad Stadium (Manchester City) as well as concerts, cricket matches and other special events.

Despite now having 99 stops, 65 miles (105 km) of track, nine different routes and two routes across the city centre, Metrolink trams do not display a route number, letter or colour, only a destination. Route colours are used on the map, however. The nine routes that operate under normal circumstances, with end-to-end journey times, are:

- Green: Altrincham–Market Street–Bury (57 minutes)
- Purple: Altrincham–Piccadilly (32 minutes)
- Yellow: Bury–Piccadilly (33 minutes)
- Orange: MediaCityUK–Ashton-under-Lyne (54 minutes)
- Turquoise: Eccles–Ashton-under-Lyne (63 minutes) (serving MediaCityUK in the evenings and on Sundays)
- Navy: Manchester Airport–2CC–Victoria (58 minutes)
- Pink: East Didsbury–2CC–Rochdale (1h18)
- Grey: East Didsbury–2CC–Shaw & Crompton (1h02)
- Red: Cornbrook–intu Trafford Centre (16 minutes).

Above: M5000 cars 3008 and 3055 pause at St Peter's Square with an Altrincham–Piccadilly service on 16 June 2019. From left to right the buildings are the Midland Hotel, Manchester Central Library, and the Town Hall extension. *Alan Yearsley*

Above: The present day Metrolink station at Victoria on 9 October 2019, with its three tracks and two island platforms (the middle track having a platform on both sides). From left to right, car 3017 waits to depart for Shaw, 3116 forms a Manchester Airport service (starting from Victoria), and 3003 is on a Bury–Piccadilly service.

Above: 3046 departs from Victoria with a Bury–Altrincham service on 8 October 2016. Behind the tram can be seen the then new overall roof of Victoria station.

Above: 3018 heads off Aytoun Street away from the camera towards Piccadilly station with a Bury–Piccadilly service on 17 April 2012. *Robert Pritchard (3)*

Above: M5000 car 3020 calls at Stretford with an Altrincham service on 17 April 2012. *Alan Yearsley*

Below: 3020 calls at Monsall, the first intermediate stop on the Rochdale line, with a St Werburgh's Road service on 19 April 2013.

Above: M5000 car 3018 calls at New Islington with a Droylsden–Bury service on 19 April 2013. *Alan Yearsley (2)*

Above: M5000 car 3077 calls at Exchange Quay with an Eccles–Piccadilly service on 24 August 2017. *Steve Hyde*

Above: M5000 car 3053 approaches Shadowmoss, the first stop after Manchester Airport, with a Deansgate-Castlefield service on 12 November 2017. *Robert Pritchard*

On Sundays (and at all times under the reduced service operating at the time of going to press due to the Covid-19 pandemic) the Green, Orange and Grey routes do not run, the Turquoise route runs via MediaCityUK using the north side of the triangle near Harbour City, and the other five routes run to the standard 12-minute frequency.

Under the reduced Covid-19 timetable in operation at the time of writing (February 2021) service frequencies across all five operational routes are:

- Monday–Friday: 06.00–19.00 every ten minutes; 19.00–00.00 every 20 minutes
- Saturday: 06.00–08.00 and 20.00–00.00 every 20 minutes; 08.00–20.00 every ten minutes
- Sunday: 07.00–23.00 every 15 minutes.

The Monday–Saturday service requires 108 of the 124 M5000 trams, both under normal circumstances and with the reduced timetable. Metrolink is currently the only UK tram network to use pairs of trams in normal service. On Mondays–Saturdays under the normal timetable all 11 diagrams on the Green route are booked for pairs of trams, as are two diagrams each on the Purple and Yellow routes and four on the Pink and Grey routes (these two routes interwork, for example a diagram will run East Didsbury–Rochdale–East Didsbury–Shaw & Crompton–East Didsbury etc). All Sunday diagrams are normally for single trams except during football or cricket matches or other major events. Under the reduced timetable in operation during the pandemic, more double trams are operated to help with social distancing. Services are kept under review and double sets are moved as changing pandemic restrictions affect levels of demand.

Above: M5000 car 3017 heads away from the camera as it departs from Clayton Hall with a Droylsden–Bury service on 19 April 2013. *Alan Yearsley*

Above: M5000 car 3001 leaves the East Didsbury terminus with a Rochdale service on 25 May 2013. *Robert Pritchard*

Above: M5000 car 3025 approaches Exchange Square with an East Didsbury service on 1 April 2017 as car 3075 (left) heads towards Victoria with a Shaw & Crompton service. The National Football Museum is just visible on the left. *Alan Yearsley*

Above: M5000 car 3041 arrives at South Chadderton with a Rochdale service on 19 April 2013.

Above: M5000 car 3099 is seen on Cross Street on the Second City Crossing shortly after departure from Exchange Square with an East Didsbury service on 14 June 2019. *Alan Yearsley (2)*

Left: M5000 car 3083 passes Barton Square as it approaches intu Trafford Centre with a shuttle from Cornbrook on 23 March 2020. *Paul Abell*

Above: M5000 car 3102 calls at Imperial War Museum with a Cornbrook–intu Trafford Centre shuttle on 18 July 2020. *Alan Yearsley*

TICKETING AND FARES

Tickets must be purchased from the ticket vending machines (TVMs) at stops before boarding the tram, except for passengers who already hold a valid ticket, pass or contactless payment card or app (see below). For the first 17 years of operation Metrolink used Thorn EMI TVMs, which issued simple paper tickets and only accepted coins but gave change if available. The original versions of these TVMs, installed at tram stops on the Phase 1 network, had a push button panel with four columns of six blue destination buttons with a column containing four yellow and one white ticket types plus a red cancel button. Tram stops were grouped into three zones, and passengers had to consult notices on or adjacent to the TVMs to find out the

Above: This view of New Islington tram stop on the Ashton line shows a standard Metrolink waiting shelter and Scheidt & Bachmann ticket vending machine.

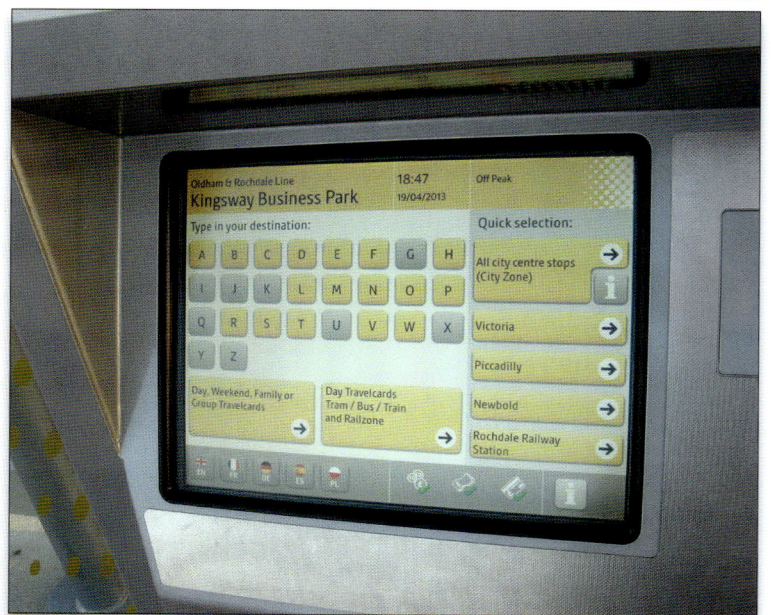

Above: A close-up of the touch-screen display on the TVM at Kingsway Business Park stop on the Rochdale line. *Alan Yearsley*

Left: A Scheidt & Bachmann ticket vending machine at Burton Road tram stop on the East Didsbury line. *Robert Pritchard (2)*

zone where their destination was located. A later version of the same design of TVM was installed at stops on the Eccles line with seven columns of 11 yellow destination buttons and a small LCD screen showing transaction stages, and some of these machines also accepted notes. In due course similar whole new front panels were fitted to the original TVMs and separate buttons were provided for each stop rather than just for each zone to make the machines easier to use.

Poor reliability of the Thorn EMI TVMs led GMPTE to appoint German company Scheidt & Bachmann in September 2008 to replace all TVMs across the Metrolink network with touch-screen "Ticket Xpress" machines similar to those in use at many stations across the national rail network. These accept credit and debit cards (with both contactless and chip-and-pin payment available) as well as coins and notes and issue credit card-sized tickets. Passengers can purchase single or return tickets to any destination on Metrolink and through tickets to some National Rail destinations within Greater Manchester as well as the usual range of day tickets. On 13 January 2019 TfGM introduced a new system of zonal fares to replace 8500 stop-to-stop fare combinations, although the old fare zones had continued to exist behind the scenes with more zones being added as the system expanded. The network is now divided into four zones:

- Zone 1: City centre bounded by Victoria, New Islington and Cornbrook
- Zone 2: Cornbrook–Parkway/Eccles/Stretford/St Werburgh's Road/Barlow Moor Road, Queens Road–Bowker Vale, Monsall–Newton Heath & Moston, Holt Town–Edge Lane
- Zone 3: Stretford–Brooklands, Parkway–intu Trafford Centre, St Werburgh's Road–East Didsbury, Barlow Moor Road–Roundthorn, Bowker Vale–Whitefield, Newton Heath & Moston–Derker, Edge Lane–Ashton-under Lyne.

- Zone 4: Brooklands–Altrincham, Roundthorn–Manchester Airport, Whitefield–Bury, Derker–Rochdale.

TABLE 1: ESTIMATED PASSENGER JOURNEYS MADE ON METROLINK PER FINANCIAL YEAR

Year	Passenger journeys (millions)	Year	Passenger journeys (millions)
1992–93	8.1	2008–09	21.1
1993–94	11.3	2009–10	19.6
1994–95	12.3	2010–11*	19.2
1995–96	12.6	2011–12	22.3
1996–97	13.4	2012–13	25
1997–98	13.8	2013–14	29.2
1998–99	13.2	2014–15	31.2
1999–2000	14.2	2015–16	34.3
2000–01	17.2	2016–17	37.8
2001–02	18.2	2017–18	41.2
2002–03	18.8	2018–19	43.7
2003–04	18.9	2019–20	44.3
2004–05	19.7		
2005–06	19.9		
2006–07	19.8		
2007–08	20		

*In 2010–11 Metrolink revised its method of calculating passenger boardings so the figures for 2010–11 are not directly comparable with previous years.

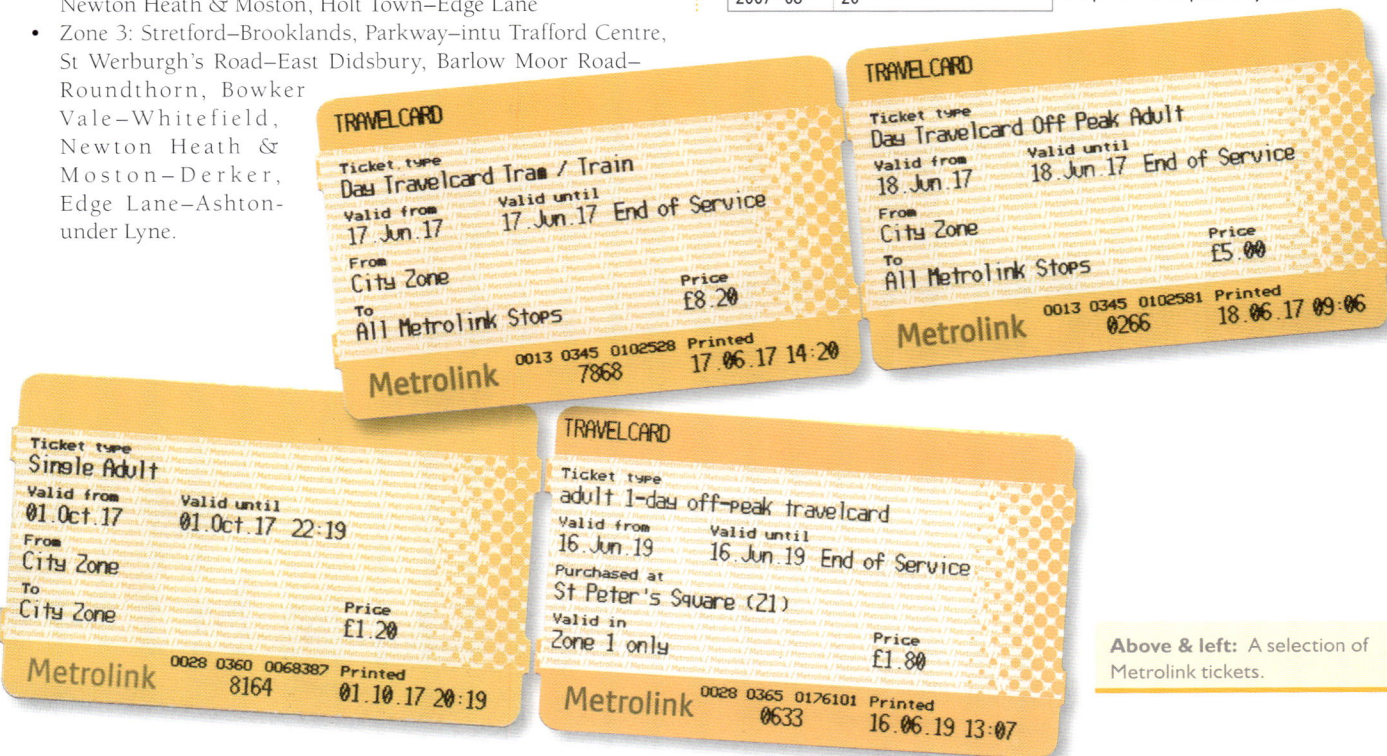

Above & left: A selection of Metrolink tickets.

A single ticket for travel within any one zone costs £1.40 at 2020 prices (fares for 2021 had not been announced at the time of going to press). An all-zones Peak Day Travelcard valid only on trams is £7.10, an Off-Peak Travelcard costs £4.90 and one for a weekend (valid after 18.00 Fridays) £6.80. "System One" tickets also cover buses and/or trains in Greater Manchester and there is also the GM Wayfarer, which for £14.40 allows off-peak travel on trams, trains and buses throughout Greater Manchester and also parts of Cheshire, Derbyshire, Lancashire, Staffordshire and the Peak District. GM Wayfarers are not available from Metrolink TVMs but can be bought from rail stations, TfGM travel shops or PayPoint outlets.

Contactless payments can also be made using TfGM's "get me there" app or smartcard or a contactless credit or debit card. This can be done by simply tapping your card or phone on the card reader at the tram stop before boarding the tram and after alighting. The system then automatically applies the appropriate daily price cap (if travelling across all four zones this is £4.90 off-peak or £7.10 peak) if enough journeys are made on the same day using the same card or device. At present, the "get me there" app and contactless payment cards can only be used for travel on Metrolink (contactless cards are also accepted on most buses but bus journeys made in this way will not count towards the daily price cap). Smartcards can also be used for "System One" bus and tram combined tickets but must be "touched in" at a tram stop before being used on a bus.

Through single and return tickets for combined tram and train journeys are also available. Rail tickets issued for travel to "Manchester CLTZ" (Manchester Central Zone) or for any journey within Greater Manchester that involves a transfer between Piccadilly and Victoria stations are valid on Metrolink within

Zone 1 only (although such tickets are generally also valid via the Ordsall Chord between Oxford Road and Victoria as an alternative whether for a journey starting and finishing inside or outside Greater Manchester). If travelling from a rail station within the TfGM area to a Metrolink stop outside Zone 1, combined tram and train single or return tickets must be bought from the ticket office or on the train if boarding at an unstaffed station. If the journey starts at a Metrolink stop, such tickets can be bought by pressing the "Combined Travel – Tram Bus Train and Railzone" button and then choosing "Railzone" and the appropriate rail zone.

RIDERSHIP FIGURES

44.3 million passenger journeys were made on Metrolink in 2019–20, compared to 43.7 million in 2018–19. In 1992–93, Metrolink's first year of operation, 8.1 million journeys were made. Almost every year since Metrolink opened in April 1992 the number of journeys has increased, with a few notable exceptions (see Table 1), reflecting the growth of the network as well as increasing demand.

ACCIDENTS

Metrolink has a very good safety record, with only a relatively small number of serious accidents having occurred in its 29-year history. Some of the most notable incidents include:

- On 12 August 1996 a lorry collided with a tram at the junction of Corporation Street and Miller Street. 16 people were injured, one seriously.
- On 18 October 2002 an 18-year-old woman was killed after slipping and falling onto the track in front of a moving tram near Manchester Central Convention Centre.
- Three derailments were caused by the failure of a rail keep on a curved section of track, at Shudehill on 31 August 2004, at London Road on 11 January 2005, and at the start of the street running section at Long Millgate just outside Victoria on 22 March 2006. A fourth derailment at Pomona on 17 January 2007 was caused by out of tolerance track gauge, which widened further as the tram travelled over it.
- On 25 June 2005 a 16-year-old man was killed when he was hit by a tram after straying onto the track at Navigation Road stop.
- On 29 June 2008 the middle bogie of a tram derailed just after departing from St Peter's Square towards Piccadilly Gardens. This incident was later found to have been caused by a track fault. A number of passengers suffered minor injuries.
- On 5 June 2011 a 67-year-old man was killed after being hit by a tram near Piccadilly Gardens.
- On 15 December 2011 a blind man was killed after being hit by a tram near St Peter's Square.
- On 6 February 2013 a woman was killed after being hit by a tram near Failsworth stop.
- On 11 January 2014 a man was killed after being hit by a tram near Market Street stop.

Above: The East Didsbury-bound platform at Burton Road tram stop, showing the waiting shelter, ticket vending machine, help point, departure indicator and smartcard/contactless payment validator. *Robert Pritchard*

- On 12 May 2015 a man was seriously injured after being hit by a tram from which he had alighted at Market Street.
- On 16 February 2016 a cyclist was killed after being hit by a tram at Robinswood Road stop.
- On 10 November 2017 cars 3044 and 3079 were in low-speed collision at St Peter's Square and sustained minor damage. There were no injuries.
- On 17 April 2019 Ray Boddington from the Manchester busking band The Piccadilly Rats was killed after being hit by a tram between Market Street and Shudehill stops.
- On 28 November 2019 a 12-year-old boy suffered minor injuries after being hit by a tram on Aytoun Street between Piccadilly Gardens and Piccadilly station.

CHAPTER 8:

FUTURE
DEVELOPMENTS

OTHER FUTURE EXTENSIONS AND TRAM-TRAIN PLANS

The Greater Manchester Transport Strategy 2040 Draft Delivery Plan 2020–2025, published in January 2019, contained a series of plans for further expansion of the Metrolink network. During the period 2020–25, TfGM plans to put together a business case for extensions of the airport line via Davenport Green and Wythenshawe Hospital (from Roundthorn) and Manchester Airport Terminal 2 and the early development of route alignments from Ashton-under-Lyne to Stalybridge, East Didsbury to Stockport, and Trafford Centre to Port Salford. The Wythenshawe loop scheme was originally proposed as part of the Manchester Airport line but was shelved in 2005 to control costs.

TfGM is also assessing the feasibility of tram-trains on Rochdale–Heywood, Manchester Airport–Wilmslow and Altrincham–Hale "pathfinder" projects, following the success of the Sheffield–Rotherham tram-train trial in South Yorkshire where the same vehicles operate on both Supertram and Network Rail metals. If these go ahead and prove successful, tram-train operation could be extended to a number of other routes including Rochdale–Heywood–Bury (using the existing East Lancashire Railway tracks), Stockport–Manchester Airport, Cornbrook–Manchester Airport via Timperley, and the existing heavy rail lines from Manchester to Glossop/Hadfield, Marple, Hazel Grove, Warrington via Trafford Park, and Wigan via Atherton. Options for these routes, with a number of new or rebuilt stations (Reddish South, Denton, White City, Timperley East, Gatley North, Adswood, Pendlebury, Dobb Brow, Little Hulton, Baguley and Cheadle), will be developed over the next few years. If tram-trains are found to be unviable, attention will turn to how to improve capacity using existing heavy rail services instead.

Other long-term projects include a new stop at Sand Hills near Queens Road to serve the Manchester Northern Gateway growth area, an extension from Bowker Vale to Middleton, a link from MediaCityUK to Salford Central, and the introduction of new, longer trams. For the time being these projects should be considered as no more than aspirations, as their feasibility is yet to be demonstrated.

Shortly before this book closed for press, Manchester and Salford City Councils, TfGM and the Greater Manchester Combined Authority published a Draft City Centre Transport Strategy covering the period to 2040 and designed to improve connectivity and reduce the use of private cars for travel into Manchester city centre. Part of this strategy

Below: The Trafford Park line could eventually be extended to Port Salford. On 23 March 2020 M5000 car 3083 arrives at Wharfside with an intu Trafford Centre–Cornbrook shuttle. Manchester United's Old Trafford football ground is adjacent to the westbound platform on the left. *Paul Abell*

Left: Could Greater Manchester be the next area of the UK to embrace the tram-train concept after the success of the Sheffield–Rotherham tram-train trial? On 3 February 2019 car 399 201 arrives at the new dedicated tram-train platforms at Rotherham Central with a Parkgate–Sheffield Cathedral service.

will involve exploring the feasibility of a Metrolink tunnel beneath the city centre if the network expands beyond the amount of space available at street level. This would also facilitate the use of longer trams. With city centre tunnels having already been mooted but never realised on a number of earlier occasions (see Chapters 1 and 3) it remains to be seen if this scheme ever comes to fruition.

Right: Class 14s D9531 and D9537 have just crossed the Metrolink line into Bury Interchange as they operate on the East Lancashire Railway's Heywood line with the 09.30 Bury–Heywood on 26 July 2014. Heywood is one of the lines that could potentially see tram-trains operating in the future.
Robert Pritchard (2)

APPENDIX: ABBREVIATIONS USED IN THIS BOOK

AC	Alternating current	GMPTE	Greater Manchester Passenger Transport Executive
AGMA	Association of Greater Manchester Authorities	LCD	Liquid-crystal display
BREL	British Rail Engineering Ltd	LRV	Light rail vehicle
CCTV	Closed circuit television	MARTIC	Manchester Area Rapid Transit Investigation Committee
DC	Direct current	MTMS	Manchester Transport Museum Society
DfT	Department for Transport	NMC	Network Management Centre
DLR	Docklands Light Railway	RATP	Régie Autonome des Transports Parisiens
DMU	Diesel multiple unit	SELNEC	South-East Lancashire North-East Cheshire
ELR	East Lancashire Railway	SPV	Special Purpose Vehicle
EMU	Electric multiple unit	TfGM	Transport for Greater Manchester
GEC	General Electric Company	TMS	Tram Management System
		TVM	Ticket vending machine